STUDENT UNIT GUIDE

UNITS

AQA AS A2 3 6

Biology

Investigative and Practical Skills in Biology

Steve Potter

The publishers would like to thank Martin Rowland for his work on this book.

First published in 2011 by Philip Allan Updates, an imprint of Hodder Education, an Hachette UK company, Market Place, Deddington, Oxfordshire OX15 0SE

Orders
Bookpoint Ltd, 130 Milton Park, Abingdon, Oxfordshire OX14 4SB
tel: 01235 827827
fax: 01235 400401
e-mail: education@bookpoint.co.uk
Lines are open 9.00 a.m.–5.00 p.m., Monday to Saturday, with a 24-hour message answering service. You can also order through the Philip Allan Updates website: www.philipallan.co.uk

ISBN 978-1-4441-1547-5

First printed 2011
Impression number 5 4 3 2 1
Year 2015 2014 2013 2012 2011

Typeset by MPS Limited, a Macmillan Company
Printed by MPG Books, Bodmin

Hachette UK's policy is to use papers that are natural, renewable and recyclable products and made from wood grown in sustainable forests. The logging and manufacturing processes are expected to conform to the environmental regulations of the country of origin.

P01865

Contents

Introduction

■ ■ ■

The Unit 3 assessment

■ ■ ■

The Unit 6 assessment

Introduction

About this guide

This guide is written to help you to prepare for the Units 3 and 6 assessments of the new AQA Biology specification. These units aim to develop your investigative and practical skills. You are not required to learn a body of facts, rather to understand the principles of how to investigate a problem experimentally. These principles include how to:

- design the investigation
- carry out the investigation
- record and present the results from the investigation
- analyse the data obtained
- draw conclusions from the data using biological knowledge that allows interpretations to be made
- evaluate the reliability and validity of the investigation

This guide takes you through the stages of carrying out a biological investigation and shows you the aspects you should consider at each stage.

It examines the issues that are often raised in assessments and how to address these in your answers.

It is not possible to consider in detail some of the specific practical investigations used in assessments, although an indication of these is given. However, the principles used in all assessments are similar and these are addressed and illustrated with examples.

How science works

This is a new component in the biology specifications of all examination boards, including AQA. The aim is to help you understand the process of scientific work. The main aspects of 'How science works' (HSW) are described below.

- Scientists use pre-existing knowledge and understanding/theories/models to suggest explanations for phenomena.
- They design, carry out, analyse and evaluate scientific investigations to test new explanations.
- They share their findings with other scientists so that they may be validated, or not, as the case may be.

As a consequence of the work of scientists, there may be implications for society as a whole. You are expected to appreciate and make informed (not emotional) comment on such aspects as:

- the ethical implications of the way in which research is carried out
- the way in which society uses science to help in decision making

How should I use this guide?

The guide lends itself to a number of uses throughout your course. You can use it:

- to help you in your design of investigations
- to help you prepare your analysis
- to help you prepare your evaluation
- to help you understand the mechanics of HSW

Preparing for the Unit 3 and Unit 6 assessments

Terms used in assessments

You will be asked precise questions in the assessments. You can save time, and ensure you score as many marks as possible, by knowing what is expected. Terms most commonly used are explained below.

- **Describe** — this means exactly what it says — 'tell me about...' — and you should not need to explain why.
- **Explain** — give biological reasons for *why* or *how* something is happening.
- **Complete** — finish off a diagram, graph, flow chart or table.
- **Draw/plot** — construct some type of graph. For this, make sure that you:
  - choose a scale that makes good use of the graph paper (if a scale is not given) and does not leave all the plots tucked away in one corner
  - plot an appropriate type of graph — if both variables are continuous variables, then a line graph is usually the most appropriate; if one is a discrete variable, then a bar chart is appropriate
  - plot carefully using a sharp pencil and draw lines accurately
- **From the**... — use only information in the diagram/graph/photograph or other forms of data.
- **Name** — give the name of a structure/molecule/organism etc.
- **Suggest** — i.e. 'give a plausible biological explanation for'; this term is often used when testing understanding of concepts in an unfamiliar situation.
- **Compare** — give similarities *and* differences between...
- **Calculate** — add, subtract, multiply, divide (do some kind of sum) and show how you got your answer — *always* show your working.

The process of science

What is science?

The word science comes from the Latin word '**scientia**', which means 'knowledge'. Science is not just about *having* knowledge: it is a system of acquiring knowledge based on scientific method. This is sometimes called **experimental science**, because it depends on experimentation to obtain the information. Experimental

science seeks to answer the question 'why?' It seeks to establish a **cause-and-effect** relationship between two factors.

One important idea in science is that any suggested explanation of a phenomenon should be *capable of being proved wrong*. If there is no way of proving it wrong, how can other people accept that it is correct? This is what distinguishes science from belief.

What is the scientific method?

This is the process by which biologists and other scientists approach their work. For centuries, people based their explanations of what they saw around them on observations alone, without testing their ideas to see if they were true. One ancient belief was that simple living organisms could come into being by **spontaneous generation** — i.e. that non-living *objects* could give rise to living *organisms*. For example, every year in the spring, the River Nile flooded areas of Egypt, leaving behind mud containing nutrients that enabled the people to grow crops. Along with the muddy soil, large numbers of frogs appeared that were not around in drier times. From this observation, people concluded that muddy soil gives rise to frogs!

Similarly, before the invention of the refrigerator, animal carcasses were hung in butcher's shops that were always full of flies. So people believed that the meat had turned into flies!

The main steps of the scientific method

The stages of the scientific method are summarised in Figure 1.

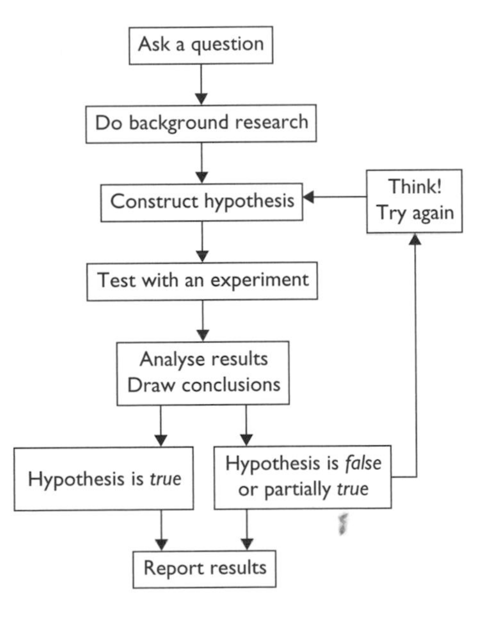

Figure 1

So what happens at each stage? What is the biologist doing? The process is centred on producing and testing a hypothesis. What is meant by 'hypothesis'?

A hypothesis is an educated guess about a possible explanation. It has to be stated in such a way that it can be tested by experiment. Let us use an example to illustrate the steps of the scientific method:

Observation: tomato seeds do not germinate inside tomatoes

Table 1

Step of the method	What happens at this step
Ask a question	A biologist knows that tomato seeds germinate when they are planted. Why don't they germinate inside tomatoes?
Do background research	The biologist now checks scientific magazines and the internet to see if this (or a similar) problem has already been researched. The biologist finds out that plants contain growth regulators.
Construct hypothesis	'There are chemicals in tomatoes that stop the seeds from growing while they are still inside the tomatoes.' This hypothesis is testable by an experiment. The biologist thinks that a chemical is responsible. How could we test that? We could try covering some seeds with tomato juice and others with water and see if any germinate. Based on the hypothesis, we can make a prediction: 'seeds covered in tomato juice will not germinate as well as seeds covered in water.'
Design and carry out an experiment to test the hypothesis	Put several tomatoes in a blender. Filter (strain) the blended material through some muslin. Collect the tomato seeds and wash them in distilled water. Place 20 seeds in a Petri dish on filter paper and cover them with the tomato juice obtained from filtering the tomatoes. Place 20 seeds in a Petri dish on filter paper and cover them with the same volume of distilled water. Place them in a growth cabinet that will keep the temperature and lighting conditions constant. Leave them for four days. Check the numbers that have germinated in each condition. Repeat the experiment 50 times.
Analyse results and draw conclusions	Out of 1000 seeds sown in each condition, 668 germinated in the distilled water (13.36 per dish); 165 germinated in the tomato juice (3.3 per dish). Something in the tomato juice is affecting germination. The cells were filtered off, so it must be a chemical in the juice.
Accept or reject the hypothesis	The hypothesis is on the right lines; the tomato juice only contains chemicals, not cells, and it reduces germination. The hypothesis is accepted. However, in the tomatoes none of the seeds germinate. There is more work to do!
Report results	The biologist must now decide whether or not to report the results to other biologists. Someone might take the work further and try to isolate the chemical responsible in order to determine what is stopping the seeds from germinating inside the tomatoes.

Disproving the idea of spontaneous generation

Remember the belief that rotting meat produces flies? In 1668, the Italian biologist Francesco Redi disproved this by using the scientific method.

Many scientists consider this to be the first true 'experiment'. Redi used wide-mouthed jars containing meat. Some jars were left open to the air. Others were covered with a piece of gauze. After several days, maggots and then flies could be seen in the open jars; none appeared in the covered jars.

Redi had *hypothesised* that only flies could produce more flies and had *predicted* that, in his experiment, flies would be found only in the open jars, He maintained all the jars under the same conditions — he *controlled many variables*. This is crucial in a scientific investigation. If there are several differences between two experimental conditions, then any one of those differences *could* account for the outcome. However, if all the variables are controlled and the difference between the experimental groups limited to the factor under investigation, then any difference in results is likely to be due to this factor alone.

By choosing to cover some jars with gauze rather than an impermeable seal, Redi allowed air to enter all the jars, which controlled a variable that could have affected the outcome of the experiment. His results matched his prediction. When other people tried the experiment, they obtained the same results. Redi concluded that flies cannot be produced by rotting meat. He also went on to say that it is unlikely that any form of spontaneous generation is possible.

The invention of the microscope opened up the world of microbiology. Many people still believed that microorganisms could arise by spontaneous generation. It took the work of Louis Pasteur to disprove this. In 1859, Pasteur carried out experiments to show that the microorganisms that caused wine and broth to go cloudy came from the air and were not made from the wine/broth itself (Figure 2).

Pasteur boiled broths in swan-necked flasks to kill any microorganisms present. The boiling forced steam and air out of the flasks. When the boiling stopped and the broth cooled, air was sucked back into the flasks. Some contained a filter to prevent solid particles from getting into the growth medium from the air. Others had no filter, but in these, the dust (and the microorganisms) in the air settled in the lowest part of the neck of the flask. All the flasks were kept under the same conditions in his laboratory.

Pasteur found that the broths stayed clear for months. At the end of this time, be treated the flasks in one of three ways:

- He left some of them as they were.
- He broke the necks on some.
- He tilted others to allow the dust in the low part of the neck to mix with the broths.

The broths in the second and third groups of flasks turned cloudy (due to the presence of microorganisms) within days. The broths in the first group remained clear.

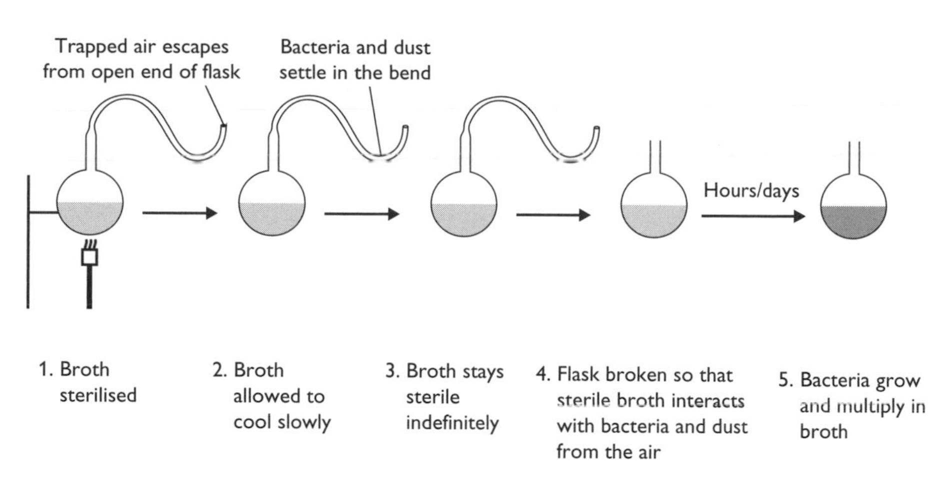

Figure 2 A summary of Pasteur's procedure

After this, people were forced to admit that spontaneous generation, even of micro-organisms, could not happen.

One stage in the scientific method is background research. Pasteur did this. He knew that other scientists had tried to disprove spontaneous generation and he drew on the results of their experiments and improved their technique.

What do we mean by cause and effect?

Scientific experiments try to establish cause and effect. This means that they try to prove that a change in one factor brings about a change in another factor. The factor that the scientist changes is called the **independent variable** (**IV**). The factor that the scientist measures to see if it changes when the IV is changed is called the **dependent variable** (**DV**). In the tomato seed investigation described earlier, the independent variable was the presence or absence of tomato juice. The dependent variable was the number of tomato seeds germinating.

To show that it *is* changes in the IV (and nothing else) that cause changes in the DV, we must take all possible steps to ensure that the experiment is a **fair test**. Other factors that could affect the results must be the same in all experimental conditions. Consider the tomato seed example. If one group of seeds had been at a higher temperature than the other group, this could have made the seeds germinate faster. We would not have known whether it was the tomato juice or the temperature affecting the results. The experiment would not be **valid**. Anything other than the IV that might influence the results must be kept constant. These are **controlled variables**. In the tomato seed investigation, the controlled variables included:

- temperature
- lighting conditions
- number of seeds per dish
- volume of liquid (water or tomato juice) added

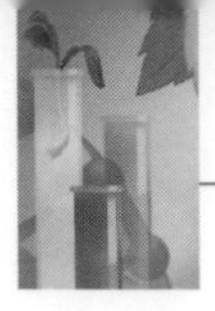

A **confounding variable** is a variable that you *can't* control and which might influence the results — it *confounds* the interpretation of the results. The presence of a confounding variable means that you cannot be certain that it is the IV that is producing the changes in the DV.

Precision, reliability and validity in scientific experiments

People often confuse these ideas. However, they are quite separate notions that are important in terms of how well an experiment is received by other scientists.

Precision (accuracy)

Precision is about how accurately you measure or count something. For example, you could measure time with a clock, a wristwatch or a stop clock accurate to 0.01 seconds. The level of accuracy chosen must reflect the magnitude of what is being measured. For example, if you are timing a reaction that is likely to last only a few minutes, a stop clock is the best choice. If you are timing something that lasts several hours, that level of precision is not needed. As a 'rule of thumb', measuring to 0.1% accuracy is enough for your purposes.

To measure volume, you could use:

- a syringe
- a measuring cylinder
- a pipette
- a burette

To obtain a volume of $2.5\,cm^3$ accurately, a pipette is best.

To measure $200\,cm^3$, use a measuring cylinder. A burette would be more precise but less convenient. Would you need the extra precision with what is quite a large volume?

Reliability

Reliability is a measure of how dependable results are. Would repeating the investigation give more or less the same results? There are several things we can do to increase reliability:

- Standardise the procedure. This makes it much more likely that we will be able to reproduce the results.
- Repeat the experiment many times. This should allow us to see a general pattern. It also allows us:
  - to spot odd results and, if justified, to exclude these
  - to calculate an average result, which is likely to be more representative than any individual result

Tip Really odd results are called anomalous results. If we are fairly sure that an error in technique caused the anomalous result, we are justified in excluding it. If we have no idea what caused it, then the result must remain. We can't exclude it just because we don't like it!

- Try not to use personal judgement. For example, if we have to wait until a solution turns a certain shade of red, one person's judgement will almost certainly differ from another person's. There are ways around this:
  - We can have a 'standard' with which to compare our experiment. In other words, something that is the exact colour needed. This helps, but we must still make a judgement.
  - We could use a **colorimeter**. This measures how much light passes through a liquid. It is nearly always better to *measure* than to *judge*.

Validity

There are two main types of validity that relate to scientific investigations:

- internal validity
- external validity

Internal validity is about whether or not an experiment measures what it says it is measuring. In the tomato seed experiment, the investigation claims to measure the effect of tomato juice on the germination of tomato seeds. We said that differences in results from the two conditions were due to the presence or absence of tomato juice. For the experiment to be valid, we must be certain that the results were due *only* to the changes in the independent variable (presence or absence of tomato juice) and to nothing else. Had we not controlled all the other variables, the experiment would not have been valid.

Tip Experiments can be reliable without necessarily being valid. If you omit the same important step consistently (perhaps by consistently forgetting to control the same variable), you may well keep getting the same results. However, the experiment will not be valid as there are at least two factors that could bring about any difference in the results.

External validity is concerned with whether or not the results of an experiment can be generalised to other situations. From the tomato seed experiment, can we conclude that chemical growth regulators prevent the germination of seeds in the fruits of, for example, different varieties of tomatoes or melons?

We would probably be fairly happy with the first idea, but may need a little more convincing about the second. It seems reasonable, but tomato plants and melon plants are very different, so we could be wrong. The external validity of our original investigation is limited.

However, if we had obtained similar results to the original investigation for seeds from apples, avocados, kiwi fruit and oranges, we would probably be happy to infer that this is also likely to be true for melons. We would have extended the investigation and could begin to see a pattern emerging that applies to all the fruits tested. We would, therefore, be more justified in concluding that it is likely to apply to all fruits. The external validity of our investigation has been increased.

A similar situation regarding external validity is found in animal experimentation to test the effectiveness of drugs. If a new drug is found to cure a condition in mice,

we could not necessarily conclude that it will have the same effect in humans. The fact that mice and humans are both mammals gives the results some external validity; the fact that humans are primates and mice are rodents limits that validity. To establish a pattern, the drug would have to be tested on more mammals. The more closely related the mammals are to humans, the greater would be the external validity of the investigations.

The Unit 3 assessment

What you must be able to do

This unit addresses the following aspects of the AS subject criteria. The ability to:

- demonstrate and describe ethical, safe and skilful practical techniques, selecting appropriate qualitative and quantitative methods
- make, record and communicate reliable and valid observations and measurements with appropriate precision and accuracy
- analyse, interpret, explain and evaluate the methodology, results and impact of your own and others' experimental and investigative activities in a variety of ways

3.3.1 Investigating biological problems involves changing a specific factor, the independent variable, and measuring the changes in the dependent variable that result.

This means that you should be able to use your knowledge and understanding of the AS specification in order to:

- identify the independent variable and describe an appropriate method of varying it
- explain how you would choose values of the independent variable to collect a full range of useful quantitative data
- identify the dependent variable and describe how you would measure it
- describe and explain the level of accuracy to which you would measure the dependent variable
- identify other variables that might affect the results, and explain what the effect might be
- describe how these variables would be kept constant
- where necessary, describe how appropriate control experiments could be established, and explain why
- distinguish between accuracy, reliability and validity and describe how you would obtain valid, accurate and reliable data

3.3.2 Implementing involves the ability to work methodically and safely, demonstrating competence in the required manipulative skills and efficiency in managing time. Raw data should be collected methodically and recorded during the course of the investigation.

This means that you should be able to use your knowledge and understanding of the AS specification in order to:

- have due regard for the wellbeing of any living organisms that are part of an investigation, and also for the environment
- carry out an investigation in a methodical and organised way, showing that you can:
  - use equipment competently

  - manage your time effectively
  - make measurements to an appropriate level of accuracy
- collect appropriate raw data and present these raw data in a suitable table

3.3.3 Raw data may require processing. Processed data should be used to plot graphs that illustrate patterns and trends from which appropriate conclusions may be drawn. Scientific knowledge from the AS specification should be used to explain these conclusions.

This means that you should be able to use your knowledge and understanding of the AS specification in order to:

- process raw data by carrying out appropriate calculations (such as calculating mean values and standard deviations)
- use the mean data to plot an appropriate graph (such as a line graph or bar chart, depending on the nature of the data)
- describe the trends and patterns in the data and, where appropriate, use specific values in the data to highlight significant changes
- recognise and distinguish between correlations and causal relationships
- draw valid conclusions, and use your biological knowledge and understanding to explain and justify these conclusions

3.3.4 Limitations are inherent in the material and apparatus used, and procedures adopted. These limitations should be identified and methods of overcoming them suggested.

This means that you should be able to use your knowledge and understanding of the AS specification in order to:

- identify aspects of the materials you use, the apparatus and your techniques and procedures that might affect the:
  - accuracy or reliability of your data
  - validity of your conclusions
- suggest *realistic* ways in which the effects of these limitations may be reduced

Unit 3 assessment: ISAs and EMPAs

There are two routes through which Unit 3 assessment can be delivered. Teachers at your centre will normally choose one of these routes for the whole centre. They are:

- **ISA (Investigative Skills Assessment)** Your teachers mark your practical assessments and submit the marks, and a selection of work as examples, to a moderator who checks the marking.
- **EMPA (Externally Marked Practical Assignment)** All the practical work you complete is marked by examiners; none of it is marked by your teachers.

Whichever route is chosen makes little difference to the nature of the tasks you have to undertake. The components of the two assessments and their approximate mark allocations are shown in Table 2.

Table 2

Assessment route			
ISA		**EMPA**	
Component	**Marks**	**Component**	**Marks**
Practical skills assessment: your teacher submits an assessment based on your practical skills throughout the course.	5	Practical skills verification: your teacher submits a verification of your practical skills, based on your work throughout the course.	0
Task Stage 1: you carry out a task specified by AQA and record your results in a table you construct. You have to decide on some aspects of the task yourself (e.g. controls, volumes, number of repeats).	3	Task 1: you carry out a task specified by AQA and record your results in a table supplied by AQA. You have to decide on some aspects of the task yourself (e.g. controls, volumes, number of repeats). You also answer some questions about aspects of the task.	10
Task Stage 2: you process the data from stage 1 in an appropriate manner (e.g. calculate means, convert volumes of gas collected into rate of collection).	6	Task 2: you carry out a second task specified by AQA and record the results in a table that you construct. You process the data in an appropriate manner and construct a graph. You answer questions on aspects of the investigation.	10
Section A of written test: you answer questions based on the results from your own investigation.	18	Section A of written test: you answer questions based on the results from your own investigation.	15
Section B of written test: you answer questions based on related material supplied by AQA.	18	Section B of written test: you answer questions based on related material supplied by AQA.	15
Total for ISA	50	Total for EMPA	50

At first glance, if your centre is following the EMPA route, you might think 'Hey, there are marks in the ISA for just doing things well all year long. I would be good at that and get 5 or 6 marks. Why aren't we doing the ISA?' You know that. Your teacher knows that. I know that and the AQA knows that. The marks for the ISA route of assessment are generally higher than those for the EMPA route. So guess what happens to the grade boundaries? Correct — the grade boundaries for the EMPA route are lower than those for the ISA route. If you want to check this for yourself, go to www.aqa.org.uk, search for 'grade boundaries', select an examination session (e.g. June 2010), the type of examination (A-level and AS (new)) and scroll down until you come to the biology grades. Bear in mind that you will only find grade boundaries for the ISAs and EMPAs in the June series.

The practical skills you will need

AQA assumes that you are familiar with the use of basic equipment, such as measuring cylinders, Bunsen burners, thermometers etc. You must be able to use the following equipment also.

The use of water baths to change or control temperature

You can use a beaker of water heated to the temperature you need, but it will cool down. There are two solutions to this.

- You can use a Bunsen burner intermittently to heat the water to just above the desired temperature and then allow it to fall to just below the desired temperature. With practice, the temperature can be maintained to within ±2°C.
- You can heat the water to a few degrees above the desired temperature and accept that it will fall to just below the desired temperature over the course of the experiment. The 'average' temperature will, hopefully, be the desired temperature. This method is less satisfactory and can only be used when there is a large volume of water that will not cool down quickly.

Better still, you could use a thermostatically controlled water bath. This should keep the temperature more constant. But bear the following in mind:

- It does not keep the temperature completely constant. It does (automatically and more precisely) what you would do with a Bunsen burner and beaker of water.
- The temperature may not be exactly what the water bath says. You should check it regularly with a thermometer.

The use of buffers to change or control pH

First, remember that pH measures the acidity or alkalinity of a solution. It measures the concentration of hydrogen ions relative to water. Solutions with a higher concentration of hydrogen ions than water have a lower pH and are acidic. Solutions with a lower hydrogen ion concentration than water have a higher pH and are alkaline. The pH scale is shown in Figure 3.

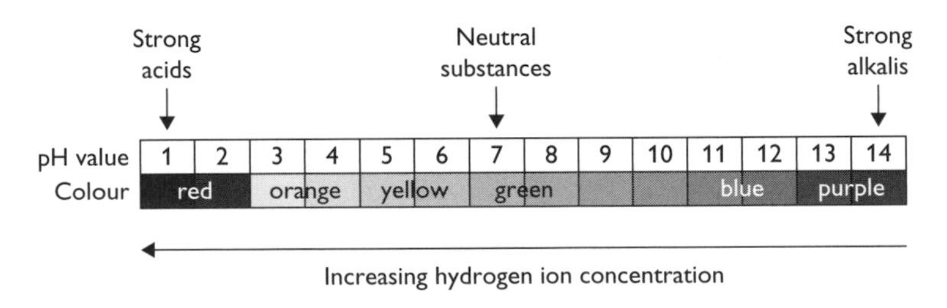

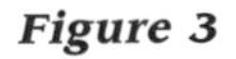

Figure 3

Adding just a tiny amount of a strong acid (or strong alkali) to a neutral solution changes its pH dramatically. Buffers are solutions that resist this change and help to maintain constant pH, within fairly narrow limits. Buffers can be prepared to maintain the pH of a solution at any given value.

Producing an appropriate dilution series when provided with stock solutions of reagents

To test the effect of changing the concentration of a solution on the rate of a process or reaction, you have to prepare the different concentrations of the solutions. There are two ways of doing this:

- You can weigh out the appropriate amount of solid for each solution and dissolve it in distilled water.
- You can repeatedly dilute a stock solution until you produce the concentration needed. This is called **serial dilution**. In your Unit 3 assessments, the stock solution will be provided.

In a serial dilution, each solution along the series is less concentrated than the previous solution by a set factor. Often this factor is 10, but it need not be. Figure 4 shows how to dilute a solution by a factor of 10.

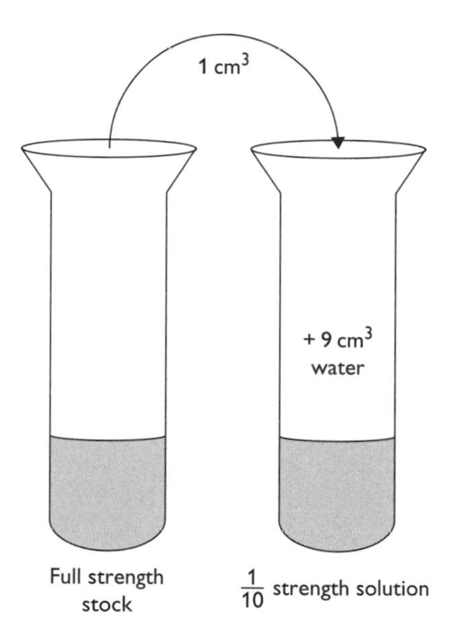

Figure 4 A 10-fold dilution of a stock solution

Notice that to get the 10-fold dilution factor, you add 9 cm^3 water, *not* 10 cm^3, to 1 cm^3 stock solution. Similarly, to get a five-fold dilution factor, you add 4 cm^3 water to 1 cm^3 stock solution.

Sometimes dilutions are made with buffer solutions, so that the pH of each dilution is the same.

Figure 5 shows how to produce two-fold, five-fold and ten-fold serial dilutions of buffer solutions.

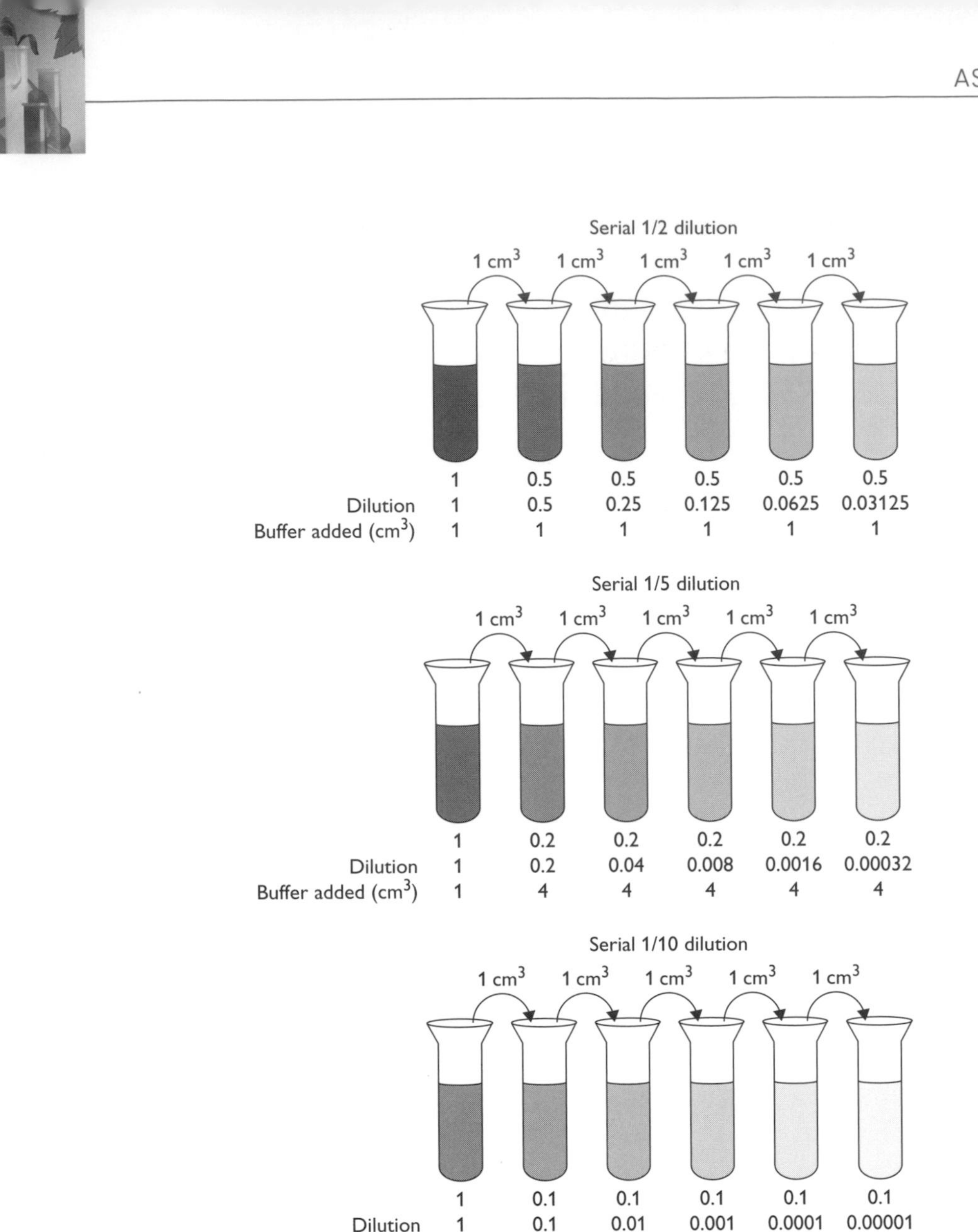

Figure 5

Using an optical microscope, preparing temporary mounts, staining and estimating size

Setting up and using an optical microscope

There is a clear sequence of steps to take to achieve perfect viewing.

1. Place the low-power objective lens in position. (In changing from one objective to another, you will hear a click when the objective is set in the correct position).
2. Turn on the substage light (or adjust the mirror if the microscope does not have one). Position the condenser as high as it will go and open the iris diaphragm by means of the lever beneath the condenser, which is below the stage of the microscope. This will give maximum light.

3 Using the coarse adjustment, raise the objective lens away from the stage. Place a properly prepared slide (see p. 20) on the stage and secure with the stage clips (or mechanical stage depending upon which is present on your microscope). Place the slide with the object directly above the brightly illuminated substage condenser.
4 Viewing the stage from the side, lower the low-power objective lens until the slide is almost (but not quite) touching it. Looking through the eyepiece, raise the objective lens slowly by turning the coarse adjustment knob counter-clockwise until the object is in focus. Use the fine adjustment to bring the object into sharp focus. If necessary, readjust the amount and intensity of light for comfortable viewing — sometimes a less intense light can give a better image.
5 To increase magnification, first, be sure your slide is focused under low power. Then, turn the nosepiece to the next highest powered objective lens. Watch from the side to make sure that the objective lens does not touch the slide. You may need to make a slight adjustment with the fine focus to bring the object into sharp focus again.

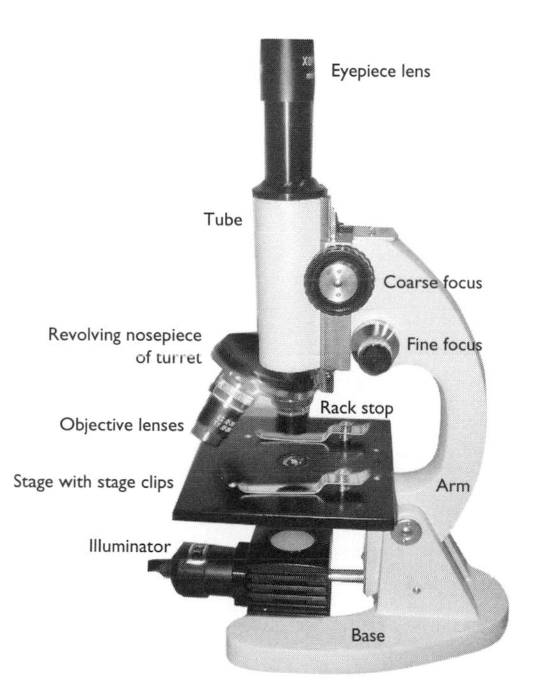

Figure 6 The main parts of a typical optical microscope

Preparing a temporary mount

To look at cells, or live specimens, you will normally prepare a wet-mount (temporary) slide, as follows:

- Place a drop of water on a clean slide with a dropper.
- Put the specimen in the water drop.
- Lower one edge of the coverslip to the edge of the water drop as shown in Figure 7. Lower the coverslip slowly to avoid trapping air bubbles in the water, as these will interfere with viewing the specimen.

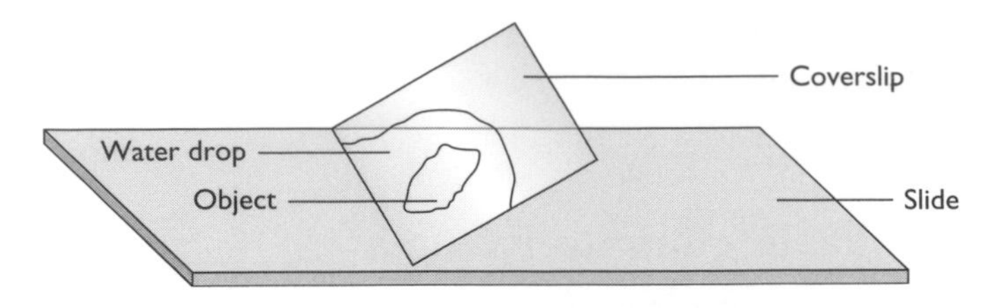

Figure 7 Preparing a temporary mount

To stain a temporary mount:

- place a small drop of the stain you are using at one edge of the cover slip
- place a small piece of paper towel or tissue paper in contact with the water *at the opposite edge* of the cover slip from the drop of stain (this will remove water from the slide, which will be replaced by the stain)

The field of view and estimating the size of an object

When you view an object under the microscope it lies inside a circular field of view. If you know the diameter of the field of view you can estimate the size of an object seen in that field. Each different magnification has a different-sized field of view. As you increase the magnification, the field of view (and diameter) gets proportionately smaller.

How to estimate the size of a cell

- Place your slide of cells (e.g. onion epidermis cells) under the microscope and get the cells in focus at magnification ×100.
- Keeping the magnification the same, remove the slide and replace it with a transparent plastic ruler.
- Focus on the millimetre scale on the ruler. You will probably see something like Figure 8.
- Use this to estimate the width of 'field of view'. You would probably estimate the field of view as shown in Figure 8 to be about 2 mm.
- Replace the slide of onion cells and re-focus.
- Count how many cells fit lengthways and widthways into the field of view.
- If eight cells fit across this field of view, the width of each cell is 2 mm ÷ 8 = 0.25 mm.

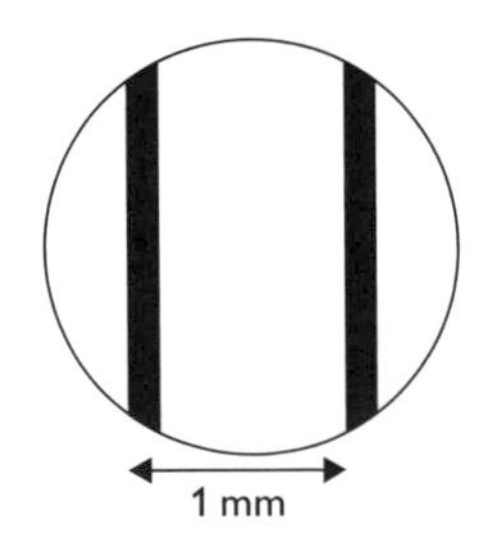

Figure 8 Estimating the field of view

This is only one estimate. You should repeat the procedure in several areas of the slide and find the average.

Remember to re-estimate the field of view if you change magnification.

Collecting reliable quantitative data

Data collection when a gas is evolved

'Evolution of gas' means the production of a gas. There are several ways in which a gas produced in a reaction can be collected, and each has plus and minus points. One way is by displacing water from a container (Figure 9).

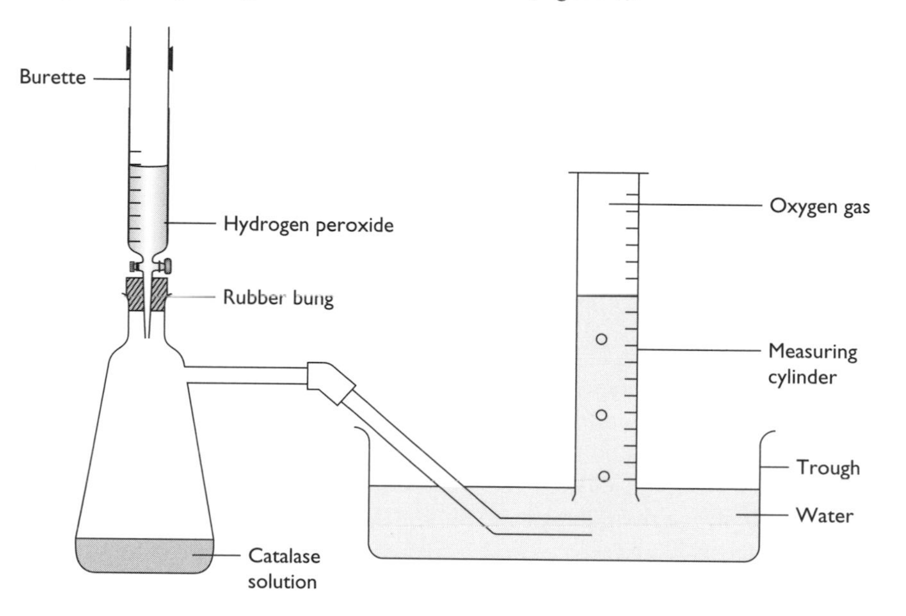

Figure 9 One method of collecting a gas by displacing water

There are a several points to note:

- A burette is used to add the hydrogen peroxide. This has a tap so everything else can be set up and the hydrogen peroxide added last. None of the gas produced will be lost.
- The burette is a graduated instrument and so a precise amount can be added.
- The measuring cylinder is a graduated instrument, so the volume of gas produced can be measured easily.

Another way is to use a gas syringe (Figure 10).

A gas syringe is a precisely made instrument. The graduations are usually $1\,cm^3$ apart, which allows you to estimate volumes to the nearest $0.5\,cm^3$. A burette is used to add the hydrogen peroxide so that none of the gas produced escapes.

There are several variations on each method — for example a syringe might be used to add the liquid to the reaction vessel. This is less precise than using a burette,

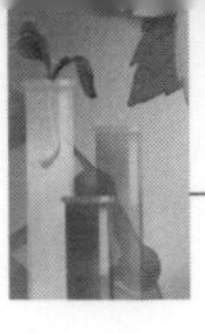

but syringes are much cheaper! In the first method, the gas could be collected in a burette for increased precision — but measuring cylinders are much cheaper!

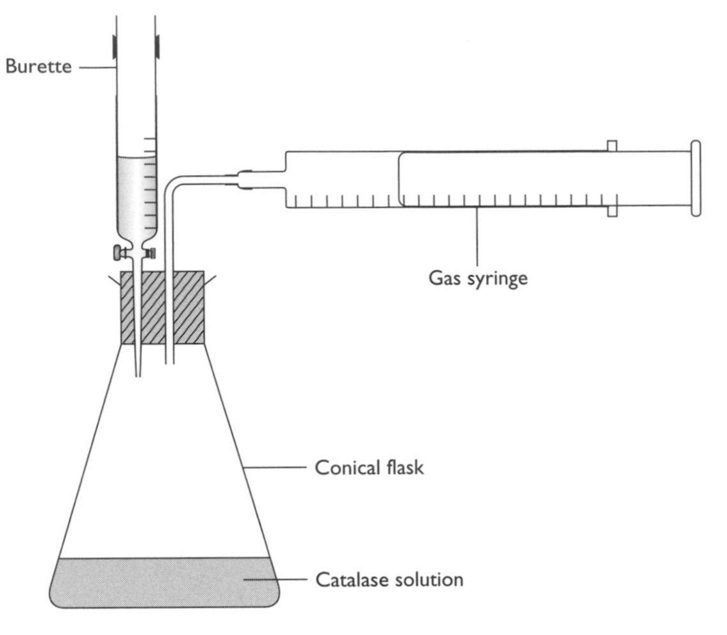

Figure 10 Using a gas syringe to collect a gas

The method chosen may be influenced by both the experiment and the number of repeats. For example, it is easier to disconnect, reset and reuse a gas syringe than it is to refill a burette or measuring cylinder with water and invert it, making sure that no water escapes.

Data collection when a colour change takes place

There are several things that you may be asked to do in investigations involving colour changes, for example:

- describe the colour change that has taken place
- use the intensity of a colour change to estimate the strength of a solution
- measure how long it takes for a specified colour change to take place

If you have to describe the colour that forms, then you should have no problem (unless you are colour-blind, in which case you should make this known to both your teacher and the AQA), provided that you can remember the basic colours of the spectrum (and black):

red, orange, yellow, green, blue, indigo, violet (and black)

Try to stick with these and not invent others. Purple/mauve probably equates to indigo and/or violet. Remember that this is a science examination and no-one is interested in artistic opinion.

If you have to time when a certain colour appears or disappears, then you have to use personal judgement. The main problem here is knowing when a certain colour change has taken place, or how much of a colour change has taken place. People's opinion about the colour of a solution varies. To get consistent judgement of a colour change you must either:

- use a colour standard for comparison
- use equipment that can *measure* the colour of the solution (such as a colorimeter or a spectrophotometer)

Using a colour standard

If you are asked to estimate the strength of a solution of glucose, using the results from the Benedict's test, you will need colour standards (Figure 11) with which to compare the results.

To produce these, you would:

- make up solutions of glucose with a range of concentrations
- test each glucose solution with Benedict's solution and retain for reference

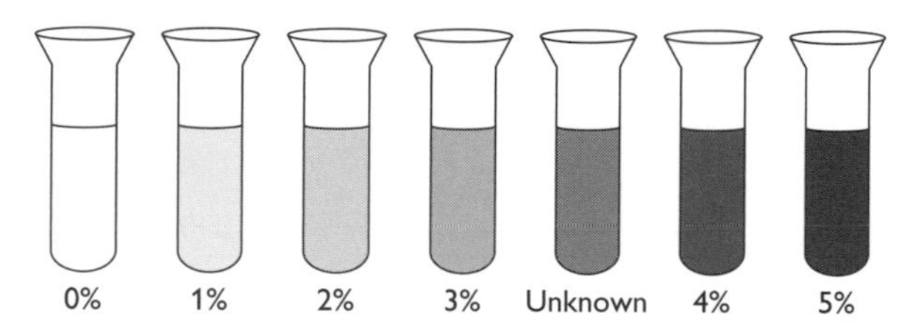

Figure 11 Preparing colour standards

To ensure that these reference colours are a valid representation (that is, they do show what your labelling says they show) you must:

- use the same volumes of reagents each time
- heat to the same temperature each time
- heat for the same length of time each time
- stop the reaction immediately time is up by adding hydrochloric acid

Test the unknown solution in exactly the same way and compare its colour with the colour standards. It may match one of them or fall between two, either way you will be able to give an estimate of its concentration. In the example shown, the concentration of the 'unknown' solution is between 3% and 4%.

Using a colorimeter

A colorimeter is a piece of equipment that measures the change in the intensity of light as it passes through a solution. Figure 12 shows what happens.

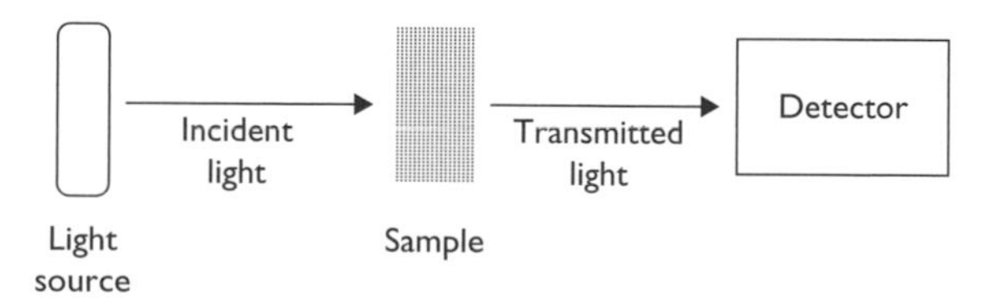

Figure 12 How a colorimeter measures the change in intensity of light passing through a solution

The colorimeter can record either:

- the amount of light that passes through (transmission) or
- the amount of light that is absorbed (absorbance)

The method is as follows:

- Make up standard solutions and test them with Benedict's solution.
- Place a small sample of each into the colorimeter and measure the transmission or absorbance.
- Plot a graph of absorbance/transmission against concentration.

The graph is called a calibration curve or calibration graph. An example of a calibration curve using absorbance (more concentrated solutions are more intensely coloured and therefore absorb more light) is given in Figure 13. Knowing the absorbance of the unknown solution, its concentration can be read from the graph. The concentration falls between 3% and 4%, but we can be more precise: the concentration is 3.7%.

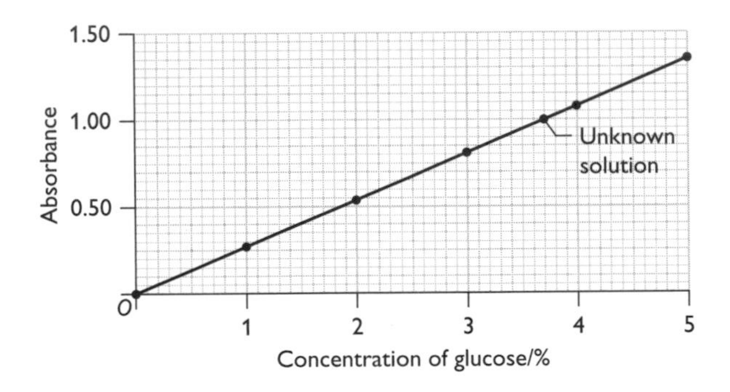

Figure 13 A calibration curve

To measure how long it takes for a colour change to take place, you could use either colour standards or a colorimeter. In either case, you would take samples periodically and compare them with the colour standards or the calibration curve.

Data collection when there are changes in mass or length

Changes in length are measured using a ruler (or millimetre graph paper because it is easier) and changes in mass using a balance. All that is called for is care and patience. Make the measurements carefully and as accurately as the equipment will allow.

Get used to recording your results in a table as you go along. You should construct your table so that is shows all the repeats at all the conditions of the investigation. It should show the original condition, the new condition and the change in the condition. It need not show mean changes or percentage changes as these are processed data and not part of the experimental results.

For example, if you measured the change in mass of potato tissue in six different concentrations of sucrose solution and carried out three repeats at each concentration, your table would look something like Table 3.

Table 3

Concentration of sucrose/g dm^{-3}	Repeat	Initial mass/g	Final mass/g	Change in mass/g
0	1			
	2			
	3			
1	1			
	2			
	3			
2	1			
	2			
	3			
3	1			
	2			
	3			
4	1			
	2			
	3			
5	1			
	2			
	3			

Increases in mass are shown as + (gain) and decreases as − (loss).

You need to be able to calculate both mean change in mass and percentage change in mass.

Mean change in mass is simple. For each concentration, add the changes, remembering to take note of the sign (some may be + and others −), and divide by the number of repeats. For this example:

$$\text{mean change} = \frac{\text{change 1} + \text{change 2} + \text{change 3}}{3}$$

Percentage change is a little more complicated, but is quite easy once you understand the principles. You are converting the actual change to *a percentage of the original.* To do this you must:

- convert the actual change to a fraction of the original
- convert the fraction to a percentage

This is done in one simple calculation, as shown below.

$$\text{percentage change} = \frac{(\text{final} - \text{original}) \times 100}{\text{original}}$$

So, the percentage change in mass of a piece of potato that increased in mass from 2.6 g to 3.12 g is:

$$\% \text{ change} = \frac{(3.12 - 2.6) \times 100}{2.6} = \frac{0.52 \times 100}{2.6} = 20\%$$

The mass has increased, so the percentage change is positive. If the mass change had been a decrease of 0.52 g, the percentage change would be −20%.

Specific investigations you need to know about

Besides the practical skills described, there are certain investigations you need to know about. It is unlikely that you will be asked to carry out the particular version of these investigations that you have carried out in your school or college. However, you may be asked to carry out, and comment on, an investigation testing the same relationship in a different context.

For example, you may have carried out an investigation into the effect of pH on the activity of the enzyme catalase in decomposing hydrogen peroxide. You would have:

- devised a way of changing the IV (independent variable — pH)
- devised a way of measuring changes in the DV (dependent variable — rate of decomposition of hydrogen peroxide by catalase)
- controlled other key variables (such as temperature, enzyme concentration, substrate concentration, volumes of the enzyme and substrate, time allowed for the reaction to proceed) in order to increase the internal validity of the experiment
- repeated the experiment a minimum of three times at each pH in order to increase the reliability of the results
- recorded your results in a table
- calculated mean reaction rates for each pH value
- plotted a graph of mean reaction rate against pH
- analysed the graph and described any patterns
- explained the patterns using biological knowledge of enzyme action
- evaluated the experiment to highlight any unavoidable errors and other limitations that were a result of the equipment used or the procedure chosen and, if possible, suggested how these errors and limitations could be minimised

It is unlikely that an ISA would ask you to carry out a practical that is used widely in schools and colleges. The main reason for this is that there will be schools and colleges that do *not* use it and so their students would be at a disadvantage. You are more likely to be asked, for example, to carry out and comment on the effect of pH on an enzyme that you have not come across before. All the aspects described above will still apply, but you will have to work out, from the information given, how to measure the dependent variable.

In this investigation, you may have used apparatus like that shown in Figure 10 to collect and measure the volume of oxygen given off during the reaction.

pH could be altered by adding different buffers to the catalase. Temperature could be investigated by standing the conical flask in a water bath at different temperatures. Using the same concentrations and volumes of catalase and hydrogen peroxide each time controls these variables.

In an ISA, you could be asked to investigate the effect of pH on any enzyme — for example pectinase. This enzyme increases fruit juice production by breaking down the pectin that holds cells together in fruits. You could use apples and investigate how much juice is produced at different pH values. You would use buffer solutions to vary the pH and control:

- the mass of apple tissue used (affects substrate concentration)
- the surface area of apple tissue
- temperature
- enzyme concentration
- volume of enzyme

You could follow this method:

- Chop apples of the same variety into 5 mm cubes.
- Weigh equal masses of chopped apple (50 g) into separate beakers.
- Prepare enzyme solutions at different pH values by adding 1 cm^3 buffer solution to 19 cm^3 enzyme solution.
- Add 5 cm^3 enzyme solution to each beaker.
- Stir the chopped apple pieces to mix all the pieces with the enzyme.
- Put both beakers into a water bath at 40°C for 20 minutes.
- Use paper coffee filters in funnels to filter the juice from the apple preparations into 100 cm^3 measuring cylinders (see Figure 14).
- When recording your results, don't forget that 5 cm^3 of the volume you collect is enzyme and buffer, so this must be subtracted from the overall volume.

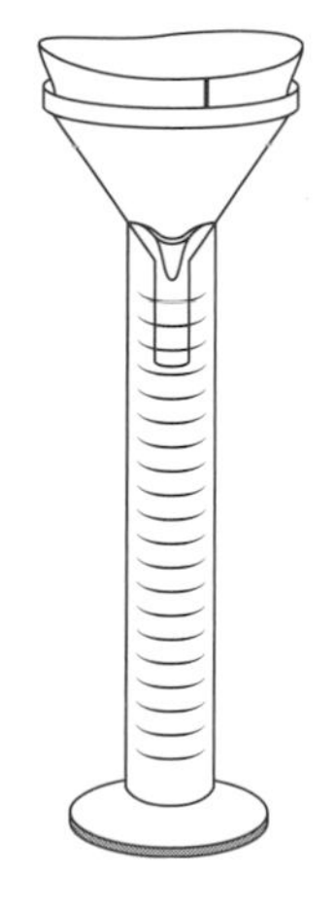

Figure 14

Even though this experiment is very different from the catalase experiment, the same principles have been used. The pH has been varied by using buffers and the other key variables have been controlled.

When the context is completely new, the ISA task sheet will tell you how to carry out the basic investigation. It will tell you how to measure the dependent variable and give a good indication of how to change the independent variable. However, you will have to decide how to control any other variable(s) and also how many repeats you will need to carry out.

Table 4 lists the experiments defined by the specification. Any one of these may be used as the basis of an ISA. Remember, the context will almost certainly be completely new to you.

Table 4

Practical	Independent variable	Dependent variable	Controlled variables
Effect of pH on enzyme activity	pH	Enzyme activity	Enzyme concentration, substrate concentration, reaction volumes, reaction time, temperature
Effect of temperature on enzyme activity	Temperature	Enzyme activity	Enzyme concentration, substrate concentration, reaction volumes, reaction time, pH
Effect of substrate concentration on enzyme activity	Substrate concentration	Enzyme activity	Enzyme concentration, pH, reaction volumes, reaction time, temperature
The three investigations above are examples of the specification requirement that 'candidates should carry out investigations into the effect of a specific variable on an enzyme-controlled reaction'.			
Effect of age on human heart rate	Age of individual	Resting heart rate	Exercise levels, gender, time of day, cigarette smoking
Effect of intensity of exercise on human heart rate	Intensity of exercise	Heart rate	Duration of exercise, type of exercise, age, gender, smoking
Effect of duration of exercise on human heart rate	Duration of exercise	Heart rate	Intensity of exercise, type of exercise, age, gender, smoking
The three investigations above are examples of the specification requirement that 'candidates should carry out investigations into the effect of a specific variable on human heart rate or pulse rate'.			
Effect of sucrose or salt solution concentration on water uptake by plant tissue	Concentration of solution	Water uptake or loss	Duration of immersion of tissue, mass/volume of tissue, surface area of tissue, absence of surface water on tissue

Practical	Independent variable	Dependent variable	Controlled variables
Effect of wind on transpiration in plants	Wind	Rate of transpiration	Duration of practical, leaf area on shoots, temperature, humidity
Effect of intra-specific competition on the abundance of a species	Extent of intra-specific competition	Abundance of species investigated	Ecological factors that might vary in the area could affect results; these cannot be controlled but can be monitored

Other skills you will need

Planning skills

Distinguishing between the independent variable and dependent variable

The factor that a scientist (you, in this case) *changes* or *manipulates* is the independent variable (IV). The factor that is measured to see if it changes when the IV is changed is the dependent variable (DV). Table 5 identifies the IV and DV in some common investigations.

Table 5

Investigation	IV	DV
The effect of pH on the activity of catalase	pH	Activity of catalase
The effect of concentration of solution on water uptake/loss by potato tissue	Concentration of solution	Rate of water uptake/loss
The effect of humidity on the rate of transpiration of willowherb shoots	Humidity of air	Rate of transpiration
The effect of cigarette smoking on resting heart rate	Number of cigarettes smoked per day	Resting heart rate
The effect of temperature on the rate of digestion of starch by amylase	Temperature	Rate of starch digestion

Even if you do not know the experiment, the wording often gives it away, for example:

The effect of X .... on Y

(This is the one I'll change — the IV)   (This is the one I'll measure — the DV)

Operationalising variables

In the investigation into the effect of pH on the activity of catalase, the changing pH in the investigation is the IV and the activity of catalase is the DV. But how do

you measure the activity of catalase? You have to restate the DV in a form in which it can be measured. This is called operationalising the variable.

How is this done? We need to look at what is happening during the reaction. Catalase is catalysing the decomposition of hydrogen peroxide (H_2O_2) into water and oxygen. We can summarise this in an equation:

$$2H_2O_2 \xrightarrow{\text{catalase}} 2H_2O + O_2$$

If catalase becomes more active, three things will happen:

- More hydrogen peroxide will be decomposed (not easy to measure).
- More water will be formed (not easy to measure).
- More oxygen will be formed (relatively easy to measure).

So, the rate of production of oxygen can be used to measure the activity of catalase. As a result, the IV and DV become:

- IV — the pH (we can control and measure this)
- DV — the rate of oxygen production

Other examples of operationalising variables are shown in Table 6.

Table 6

Unoperationalised variable	Operationalised variable
Rate of transpiration of willowherb shoots	Rate of water loss/uptake by willowherb shoots
Rate of osmosis in potato tissue	Rate of water uptake/loss by potato tissue
Heart rate	Pulse rate
Amylase activity	Time taken for starch to be digested

Controlling key variables

In the introduction, we discussed the need for internal validity in investigations (in other words, we need to make them fair tests). To do this, we must change only one variable at a time; the others must remain as constant as possible. Table 7 lists the key variables that affect a number of biological processes. If you carry out an investigation into one of these, you must control all the others.

Table 7

Biological process	Key variables	How controlled
Enzyme-controlled reactions	Temperature	Water bath
	pH	Buffer solutions
	Substrate concentration	Same concentration and volume of substrate for all repeats
	Enzyme concentration	Same concentration and volume of enzyme for all repeats
	Extent of reaction	Same time for all repeats

Biological process	Key variables	How controlled
Transpiration	Temperature	Could use an environmental chamber to maintain or to vary this, but often difficult to control — ensure that room temperature is as constant as possible
	Wind	Difficult to control, but can be varied using a fan
	Humidity	Enclosing plant in a bell jar or polythene bag creates a humid environment; placing silica gel (to absorb water vapour) in the container creates a dry environment
	Leaf area	Use plants with similar total leaf areas; express results as rate per cm^2 for comparability
Osmosis in plant tissue	Water potential of the plant tissue	Use the same source (e.g. the same potato) for all repeats; if this is not possible, make sure the different tissue is used evenly in all repeats
	Water potential of the external solution	Using the same volume of the solution is less important than using the same concentration
	Temperature	Water bath
	Surface area of tissue	Use tissue with the same dimensions and shape
Heart rate	Age	Make sure all participants are within same age range
	Smoking	The extent of smoking must be similar
	Exercise	Keep type and duration of exercise constant
	Gender	Use either male or female volunteers
Ecological investigations	Light intensity Temperature Soil pH Soil water content Wind speed	It is not possible to control these, but they can be monitored using appropriate equipment. However, you must be careful not to take an unrepresentative reading. For example, an area in shade will give a low reading on the light meter; this may be different later in the day and you should take more than one reading to obtain an average and so give a more representative recording.

Control experiments

On investigating the effect of one variable on another, we need to know that it is the changes in the IV that are causing the changes in the DV. The usual way that scientists go about this is to set up a **control experiment**. The basic principle is:

- If you want to be sure that a particular factor is producing a particular effect, try the experiment without that factor and see what happens!

So, if we want to be sure that the IV is causing the change in the DV, we should have a control experiment without the IV. Let's go back to the catalase investigation (see Figure 10).

We assume that it is the catalase in the conical flask that causes the decomposition of hydrogen peroxide and the release of oxygen. But we might be wrong — it could be happening spontaneously. A control experiment in which distilled water replaces the catalase solution would settle the matter. The results from such an investigation are shown in Figure 15.

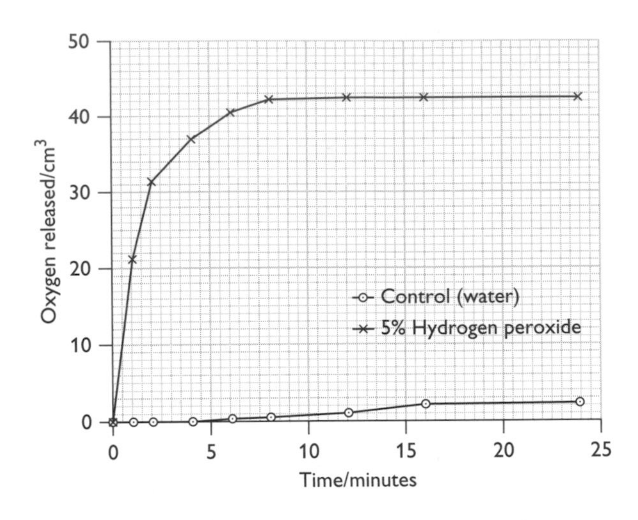

Figure 15 Results from a control experiment

The control experiment shows that there is a small amount of oxygen released, but much less than in the experimental flask. Catalase is causing the difference. Depending on the time (see Figure 15), we should subtract an appropriate amount from the volume of oxygen released to account for the control results. If investigating the effect of temperature on the activity of catalase, then, ideally, we should carry out a control experiment at each temperature.

Some further examples of the use of controls are shown in Table 8.

Table 8

Investigation	IV investigated	Control investigation(s)
Effect of substrate concentration on the activity of an enzyme	Concentration of substrate	Use distilled water instead of: • substrate • enzyme
Effect of temperature on the rate of digestion of starch by amylase	Temperature	Use distilled water instead of enzyme at each temperature
Effectiveness of different formulations of a drug	Drug and its formulation	Use placebo
Effect of intensity of exercise on heart rate	Intensity of exercise	Measure resting heart rate
Effect of wind speed on the rate of transpiration	Wind speed	Measure rate of transpiration in still air

Time management

In practical assessments, candidates often begin to carry out the tasks as soon as their teacher says that the assessment has started. They start to perform the first part of the investigation and sometime during or at the end of this they wonder how to go about the next stage. This is not a recipe for success. You need to spend some time at the start of the assessment organising your time.

The task sheet will tell you how to carry out the basic experiment. You must then decide on:

- how to control the key variables
- how many repeats to carry out

You should think about the following:

- How long is each trial going to take?
- Will I have enough time to carry out repeats one after the other, or must I do them at the same time?
- If I do them at the same time, what must I do to make sure that I do not run some repeats for longer than others?
- Can I do anything to get the next set of repeats ready while the current set is running?
- What will I have to write down in my results table?

This last point is important, because it dictates the form that your table will take. There are two possibilities:

- You will have to record a single figure for each trial (Table 9), such as the absorbance by a colorimeter, or the time taken for a certain colour to appear/disappear.

Table 9

	Dependent variable/units			
Independent variable/units	**Repeat 1**	**Repeat 2**	**Repeat 3**	**Mean**
Condition 1				
Condition 2				
Condition 3 etc.				

Although there is a column for 'mean' here, there does not have to be. The mean value is not a part of the results, it is part of processed data.

- You will have to record more than one figure per trial (Table 10) to get the actual figure you need — for example, to record by how much a sample of potato has changed length, you will have to record the initial length and the final length.

Table 10

Independent variable/units	Repeat	Initial condition of DV/units	Final condition of DV/units	Change in condition of DV/units
Condition 1	1			
	2			
	3			
Condition 2	1			
	2			
	3			
Condition 3 etc.	1			
	2			
	3			

Whichever form of table you use, make sure that you *construct it at the start of the investigation.* It does not matter that it is not the neatest table ever, but make sure that you keep to the following conventions:

- The headings must describe fully the condition you are changing/measuring.
- The different conditions for the IV should appear in the first column.
- The conditions for the DV should appear in subsequent columns.
- The units in which you measure the IV and the DV should appear in the column headings, *not* beside the figures in the body of the table.
- The body of the table should contain numbers only, not units.

Analysing skills

You have obtained the raw data (your results), now you have to analyse them. The first step in this analysis is usually processing the data.

Processing data

Why do we need to process the data? Surely, the results tell us what we want to know?

They probably do — but a research biologist may have to sift through pages of data and trends or patterns may not be apparent immediately. It can help us to see these trends if we can:

- reduce the number of figures we have to look at
- present the data in a way that shows the trends and also gives some idea of the variability of the data

Mean and standard deviation

One way to reduce the number of figures we need to consider is to calculate the mean value for each set of data (for each set of repeats for a certain condition of the IV). Calculating the mean is quite straightforward:

$$\text{mean} = \frac{\text{sum of all the values for a set of repeats}}{\text{number of repeats in the set}}$$

Think back to the catalase experiment (catalase catalyses the decomposition of hydrogen peroxide into oxygen and water). The method of measuring the activity of the catalase is to collect the oxygen produced. Table 11 shows how long it took to collect a set volume ($30\,cm^3$) of oxygen at different temperatures.

Table 11

	Time taken to collect 30 cm³ oxygen/s		
Temperature/°C	**Trial 1**	**Trial 2**	**Trial 3**
10	54	47	43
20	12	14	16
30	5	5	5
35	9	5	4
40	9	6	9
45	14	11	11
50	73	71	57
55	119	109	132

If we look at the figures in Table 11, it seems fairly obvious that it takes considerably less time at 30°C, 35°C and 40°C than at the other temperatures. It also seems clear that 35°C is the temperature at which the reaction occurs the fastest. However, the mean results (Table 12) show that there is little difference between the means for 30°C and 35°C.

Table 12

Temperature/°C	Time taken to collect 30 cm^3 oxygen/s			
	Trial 1	Trial 2	Trial 3	Mean
10	54	47	43	48.0
20	12	14	16	14.0
30	5	5	5	5.0
35	9	5	4	6.0
40	9	6	9	8.0
45	14	11	11	12.0
50	73	71	57	67.0
55	119	109	132	120.0

The mean shows the pattern in the results (Table 13).

Table 13

Temperature/°C	Mean time to collect 30 cm^3 oxygen/s
10	48.0
20	14.0
30	5.0
35	6.0
40	8.0
45	12.0
50	67.0
55	120.0

The means (not raw data) are used to plot any graphs.

Calculating the mean reduces the number of figures to be dealt with and shows overall trends more easily. However, calculating the mean can hide patterns in the data, particularly where more than one variable is at work. Look at Table 14, which gives some information about the oxygen consumption of chimpanzees walking on two legs and walking on four legs.

The means suggest that there is little difference in the oxygen consumption for the two styles of walking. However, there are considerable differences between the individuals. Younger chimpanzees use more oxygen when walking on two legs whereas older chimpanzees use more oxygen when walking on four legs. We must be selective about using mean values.

Table 14

Individual tested	Oxygen intake/$cm^3\,kg^{-1}\,min^{-1}$	
	Walking on two legs	Walking on four legs
Chimpanzee, 6-year-old male	0.28	0.18
Chimpanzee, 9-year-old male	0.26	0.18
Chimpanzee, 27-year-old male	0.15	0.21
Chimpanzee, 33-year-old female	0.16	0.29
Mean	0.21	0.22

Some of the variability in the data is removed on calculating a mean — after all, that is the whole purpose of doing the calculation. However, we can show something of the variability of the data by calculating the standard deviation. This gives a spread within which 68% of values fall (34% above the mean and 34% below the mean). The reliability of the standard deviation increases with the number of data points included. With only three data points for each temperature in the catalase–hydrogen peroxide investigation, the standard deviation is less reliable than if we had carried out more repeats. Nonetheless, it can be calculated and it is included in Table 15. You are not required to know the formula for calculating standard deviation, merely to punch the raw data into your calculator and press the 'SD' button to obtain the answer.

Table 15

Temperature/°C	Mean time to collect 30 cm^3 oxygen/s	Standard deviation
10	48.0	7.2
20	14.0	2.0
30	5.0	0.0
35	6.0	2.6
40	8.0	1.7
45	12.0	1.7
50	67.0	8.7
55	120.0	11.5

This gives some idea of the variability of the data. But, does a bigger standard deviation mean more variability? Not always. If the numbers that make up the raw data are larger, you would not be surprised if they were a little more variable than if they were smaller. So which is more variable, a mean of 67.0 with a standard deviation of 8.7 (at 50°C) or a mean of 48.0 with a standard deviation of 7.2 (at 10°C)?

One way of looking at this is to express the standard deviation as a percentage of the mean:

At 10°C: $\frac{7.2 \times 100}{48} = 15\%$;　　At 50°C: $\frac{8.7 \times 100}{67} = 12.3\%$

This suggests that even though the actual standard deviation at 50°C is bigger, it probably represents slightly less variability than the smaller standard deviation at 10°C.

Another way of assessing the variability of the data is to work out the range. This is the difference between the highest result and the lowest result in a data set. It is less useful than the standard deviation because:

- an extreme value at either end of the range will give a distorted picture of the variability of the data; such a value is termed an anomalous result
- the standard deviation uses every result obtained in its calculation, so is more representative

Anomalous results

Look at the results in Table 16. They are from an investigation into the permeability of beetroot membranes at different temperatures. The amount of purple dye leaking out at each temperature is estimated by measuring the percentage of light that passes through (is transmitted through) a sample of water in which the beetroot has been immersed for a set time. Are any of these results anomalous?

Table 16

	Colorimeter reading/% transmission			
Temperature/°C	**Trial 1**	**Trial 2**	**Trial 3**	**Mean**
0	100.0	98.5	99.0	99.2
22	93.9	95.0	96.0	95.0
42	80.1	77.0	76.9	78.0
63	26.3	29.9	31.0	29.1
81	0.7	0.9	1.1	0.9
94	0.0	0.2	0.2	0.1

There are two ways of spotting anomalous results:

- Results should be similar within a data set (all the results for a certain temperature should be similar). To check this, read across each row.
- Results should follow the overall trend (the trend here seems to be decreasing transmission with increasing temperature). To check this, look down the columns.

The set of results in Table 16 seems to conform to both these demands and we should have no hesitation in using all the results in each data set to calculate the mean.

Now look at the set of results in Table 17. Are any of these anomalous?

At first glance, it seems that most of the results in trial 3 (with the exceptions of those at 81°C and 94°C) are very different from those of the other two trials. But are they anomalous? They certainly affect the mean. The mean of all three results is shown first in the 'mean' column and the mean of just trials 1 and 2 is shown in brackets.

Table 17

Temperature/°C	Colorimeter reading/% transmission			
	Trial 1	Trial 2	Trial 3	Mean
0	100.0	98.5	84.1	94.2 (99.3)
22	93.9	95.0	82.6	90.5 (94.5)
42	80.1	77.0	69.3	75.5 (78.6)
63	26.3	29.9	19.8	25.3 (28.1)
81	0.7	0.9	0.4	0.7 (0.8)
94	0.0	0.2	0.1	0.1 (0.1)

However, the results from trial 3 show the same trend with temperature as those from the other two trials. Suppose we realise that we used a different beetroot for trial 3. We now know why the results are different from the other two trials. As the trend is consistent with the trends from trials 1 and 2, we are probably justified in including the results from trial 3 in the mean.

Look at Table 18. Are there any anomalous results and, if so, should we use them when calculating the means?

Table 18

Temperature/°C	Colorimeter reading/% transmission			
	Trial 1	Trial 2	Trial 3	Mean
0	100.0	98.5	99.0	99.2
22	93.9	95.0	75.0	88.0 (94.5)
42	80.1	77.0	76.9	78.0
63	46.3	29.9	31.0	35.7 (30.5)
81	0.7	0.9	1.1	0.9
94	0.0	0.2	0.2	0.1

The two results highlighted are not part of a trend and they are quite different from the other results in the data set. The anomaly at 22°C for trial 3 breaks the trend of temperature–transmission whereas that at 63°C for trial 1 does not. They do affect the mean, but not so much that the overall trend is disrupted.

So what should we do? Unless we can account for these anomalous results (possibly due to some fault in technique or equipment) we are stuck with them. We cannot discard them just because we don't like the look of them! We certainly cannot discard them just because they are the highest or lowest values in the data set. Suppose we had carried out five repeats at each temperature and for 22°C the colorimeter readings had been:

95.0 91.7 88.4 85.1 82.6

There is quite a difference between highest and lowest values and, on this basis, you might think that you should exclude these two values. However, the spread of the values is fairly even, so what justification is there for deciding that these two results are anomalous? Just being the highest and lowest is not sufficient reason. So you must include them, unless you decide that there is just too much variability and repeat all five trials.

Drawing graphs and charts

Graphs are pictorial ways of representing data so that trends become more apparent. The type of graph (or chart) drawn depends on the type of data. To obtain results in biological investigations we usually measure or count something. We record such things as:

- how fast?
- how many?
- how much transmission?
- what volume?

Therefore, we end up with a number for the DV in each repeat of each condition of the investigation. What really determines the type of graph drawn is the nature of the IV. The IV could be either:

- discontinuous or categoric — there are two or more separate conditions for the IV and results are recorded for the DV in each of these separate conditions
- continuous — there are several conditions of the IV that form part of a continuous range of possible conditions (for example, five different temperatures between 20°C and 70°C are a sample of a continuous range of temperatures)

There are three main types of graph that you will draw in AS biology:

- **line graphs** — plotted between two continuous variables when there is likely to be a causal link between the IV and the DV
- **bar charts** — plotted when the IV is categoric and the DV is continuously variable
- **scattergrams** — plotted when both variables are continuous but a causal link cannot be assumed (usually due to lack of control of other variables)

This is summarised in Table 19.

Table 19

Investigation	IV data type	DV data type	Type of graph
Effect of temperature on release of oxygen from hydrogen peroxide by catalase	Continuous (temperature)	Continuous (volume of oxygen)	Line graph
Frequency of blood group A in age groups 15–20 and 35–40	Discontinuous (two separate ages)	Continuous (a percentage)	Bar chart
Effect of soil pH on the abundance of clover	Continuous (pH value)	Continuous (abundance of clover)	Scattergram
Effect of water potential of surrounding solution on water gain/loss in potato tissue	Continuous (water potential of solution)	Continuous (change in mass)	Line graph (note, may have to show + and −)
Effect of intensity of exercise on heart rate	Continuous (intensity of exercise)	Continuous (heart rate)	Line graph

The independent variable is plotted on the *x*-axis of the graph and the dependent variable on the *y*-axis.

Mean values, when calculated, are used in plotting graphs. It may be helpful to also calculate standard deviation and to show this on the graph to give some idea of the variability (and therefore reliability) of the data. Figure 16 is a **scattergram** showing how the temperature of the surface of the eyeball and the temperature inside the rectum change as the external temperature changes.

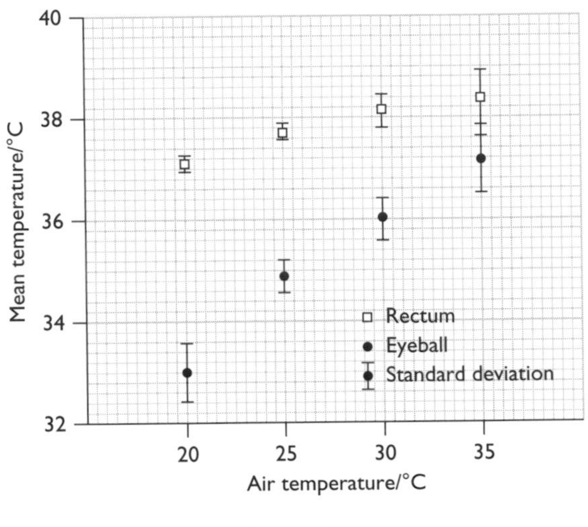

Figure 16

Notice that:

- the temperature of the eyeball changes more than that of the rectum as the air temperature changes
- the temperature of the surface of the eyeball is generally more variable than that of the rectum — this is shown by the larger bars for the standard deviation
- the temperatures of the two are very different at all air temperatures, apart from 35°C, when there is no statistical difference

How can we say that? The eyeball plot is at about 37°C whereas that for the rectum is at about 38.4°C. Remember that we are dealing with mean data and notice that the bars for the two standard deviations overlap. The standard deviations are quite large, and there is, therefore, quite a lot of variability in the data, which makes them less reliable.

If the standard deviation bars do not overlap (they don't for the other air temperatures) then, statistically, we can be confident that the two plots represent different values. However, when the standard deviation bars do overlap (as they do at 35°C) then, statistically, we cannot be confident that the two plots represent different values.

At AS you are not expected to be able to explain this in levels of probability. You just need to know that overlapping standard deviations prevent us from being confident that two mean values represent truly different things. In fact, it would be better to use a different statistic (standard error) to decide this.

Notice also that the points are not joined. This is because the graph shows the relationship between two naturally occurring variables. Other variables are not controlled, so we cannot attribute cause and effect to the situation and so we do not connect the plots. They are left as a scattergram.

Figure 17 is a **bar chart**. The independent variable is discontinuous — four species of tree.

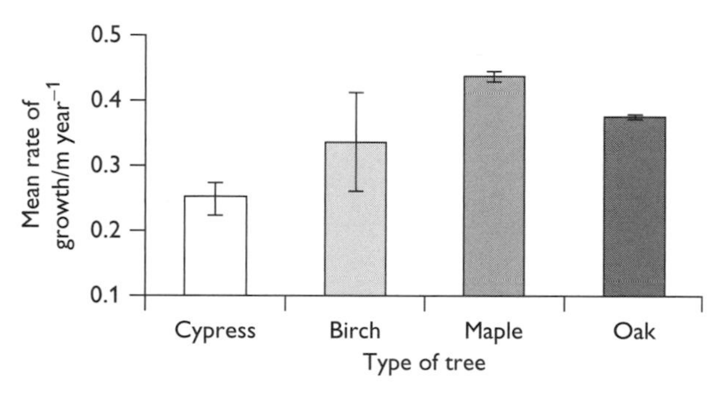

Figure 17

The dependent variable (mean rate of growth) is continuous, so we can show bars for the standard deviations as well as plotting bars for the actual rates. By comparing pairs of trees you can work out whether the growth rates are statistically different or not. For example, the standard deviations of birch and oak overlap (so are not statistically different) whereas those of oak and maple do not (and so are statistically different).

Figure 18 is a **line graph**. It shows the effect of pH on the activity of urease, an enzyme that catalyses the conversion of urea into ammonia.

Note that in this graph, the plots are joined by ruled lines. A smooth 'curve of best fit' is not drawn. There is no hard-and-fast mathematical rule about when to draw a curve (line) of best fit, and when to join plots with ruled lines. However, in this instance there are several gaps in the data. There are no results between pH 4.5 and pH 5.0. When intervening values (values between two plots) are uncertain, we should join with ruled lines. If we had both those results and some others (such as values for pH 3.7, pH 5.8 and pH 6.2) we could be more certain that all the plots are likely to be on a curve and we could draw a curve of best fit. The more data points there are, the more likely you are to be justified in drawing a curve of best fit.

Mathematicians refer to any line on a graph as a curve, whether it actually curves or is perfectly straight!

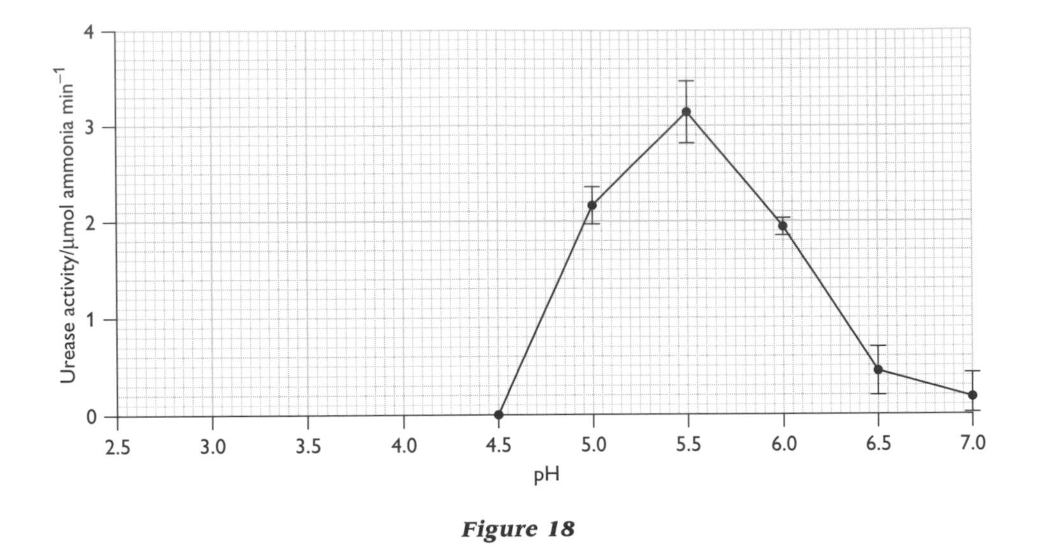

Figure 18

Being able to take a reading from a point on a line joining two plots is called **interpolation**. This is one of the values of a graph. Interpolation allows estimation of the DV at levels of the IV that have not been tested. **Extrapolation**, however, is a different matter. Extrapolation means extending the curve (which may be a straight line) of a graph assuming no change from the last known slope of the curve. There is no other reference point, so you have to be pretty sure of the data to extrapolate. In particular, 'extrapolation to zero' is a risky business — you should be careful with this.

Look at Figure 19, which shows the effect of substrate concentration on the activity of the enzyme urease.

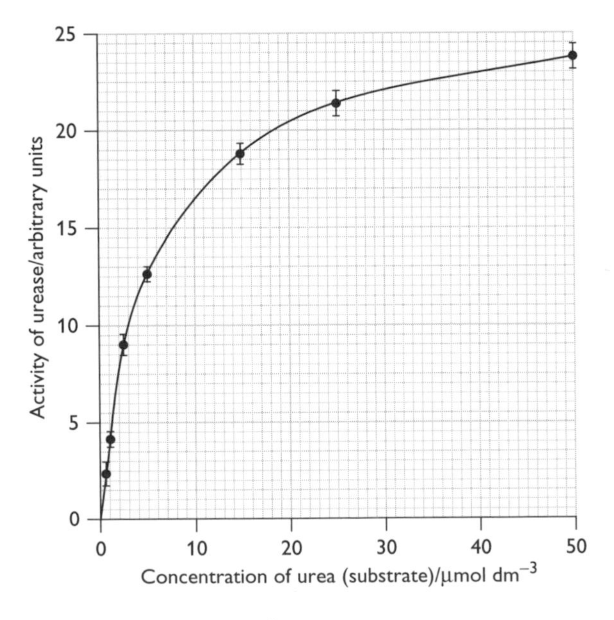

Figure 19

Although there is no plot for 0,0 (activity, substrate concentration), the curve has been extrapolated to zero. This is a reasonable assumption as, if there is no substrate, the enzyme cannot be active. However, other situations are less convincing.

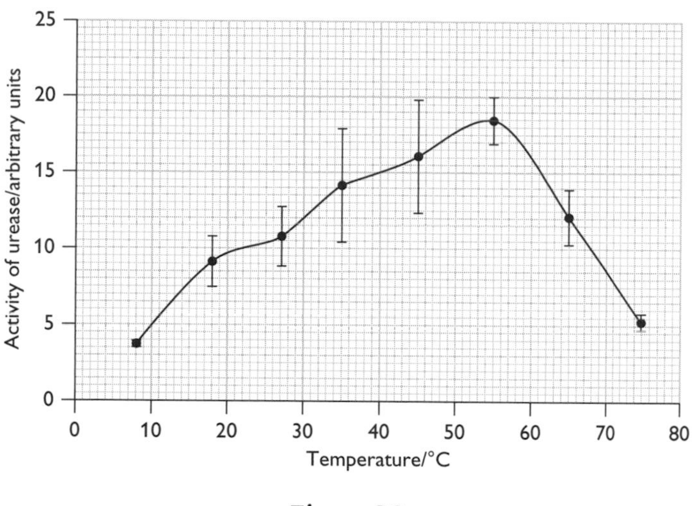

Figure 20

In Figure 20 there is a different situation. Again, there is no plot for 0,0, but there is no extrapolation to zero. This is because 0°C is not a genuine zero. As you know from cold winter days, there are temperatures lower than 0°C. Therefore, you cannot assume that at 0°C there is no activity in the same way that you could for zero substrate concentration. If you have any doubt at all about extrapolating to zero — *don't do it!*

When you are plotting graphs, make sure that you:

- choose the appropriate type of graph
- plot the IV on the *x*-axis and the DV on the *y*-axis
- choose scales that will give a large clear curve, not one that is tucked away in one corner of the graph paper

For a line graph, you should do the following:

- Use a truly linear scale on both axes.
- Plot the points accurately:
  - use small crosses or small dots with a circle around them
  - draw them with a sharp pencil
- Join the plots accurately with ruled lines or draw a curve of best fit through the points using a sharp pencil.

For a bar chart, you should do the following:

- Plot the bars accurately and draw them with a sharp pencil.
- Make sure that the bars do not touch each other.

What do we mean by a 'truly linear scale'?

Look carefully at the *x*-axis of the urease graph in Figure 18. The pH values are spread evenly along the *x*-axis. A certain distance along the *x*-axis always corresponds to the same change in pH. This is a linear scale. Now look at the version of the graph below.

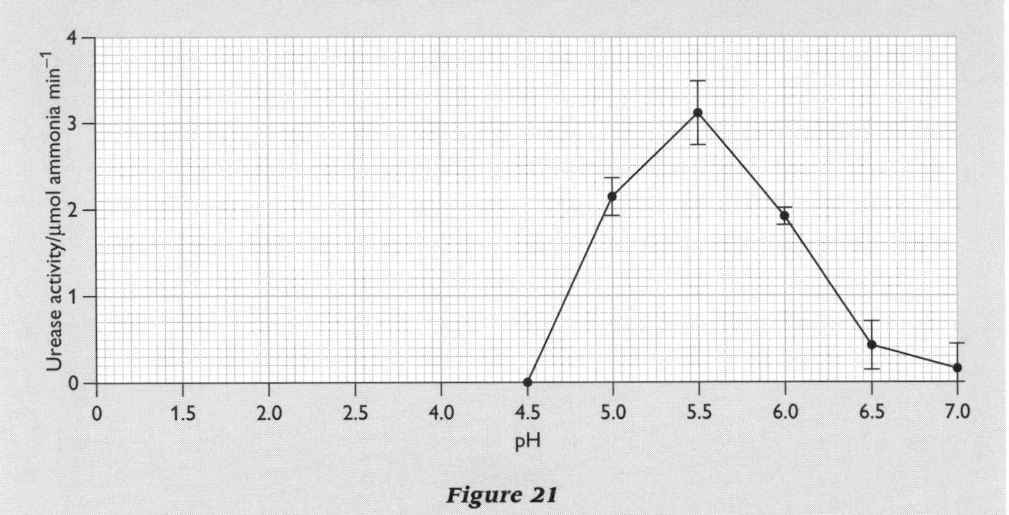

Figure 21

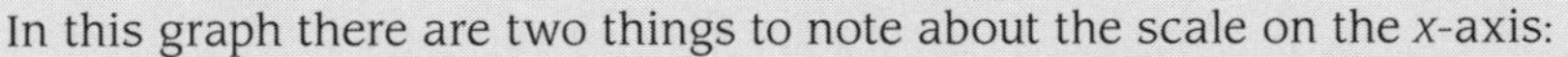

In this graph there are two things to note about the scale on the *x*-axis:

- It is not linear — the same distance can represent different changes in pH, for example:
  - the first 'unit' of distance corresponds to a change in pH from 0 to 1.5
  - the other 'units' correspond to a change in pH of 0.5, except that following pH 2.5, which again corresponds to a change in pH of 1.5
- The scale chosen results in a lot of wasted space on the graph; you do not have to start graphs at 0, 0. The *x*-axis could have been started at 2.5, 3.0 or even 3.5, giving more space to plot the graph and make trends clearer.

Interpreting and explaining data

Once graphs or charts have been drawn, we are better placed to describe any trends or patterns in the data. This should be done accurately and fully. We then use biological knowledge to try to explain these trends. Figure 22 shows the effect of substrate concentration on enzyme activity.

What is the best way to describe the trends shown in Figure 22? In describing a trend, we do not re-state every result — they should be recorded in the results table. However, it can be useful to quote one or two figures to help the description, for example:

- As the substrate concentration increases from 0 to 0.02 arbitrary units, the rate of enzyme activity increases rapidly.
- From 0.02 a.u. to 0.1 a.u., the rate of enzyme activity still increases quite quickly, but the rate of increase is slowing.

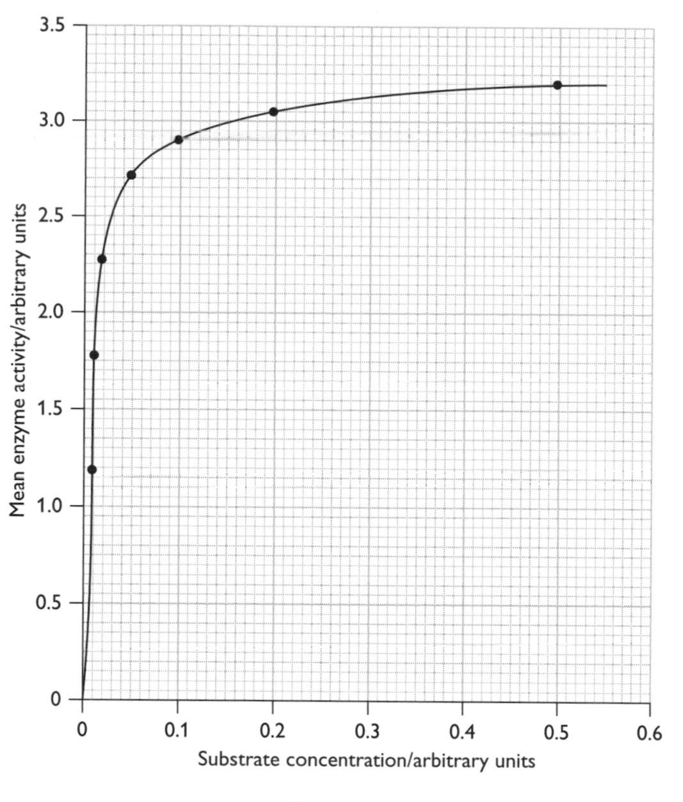

Figure 22

Tip It is important that you word this point carefully; don't say the rate of enzyme activity is slowing down — it isn't — it's still getting faster. It's just not getting faster as quickly as it was before.

- From 0.1 to 0.5 a.u., the enzyme activity increases only slightly; it is almost constant.

Having established the trend, we can use our biological knowledge of enzyme activity to explain it:

- If there is no substrate there can be no enzyme–substrate complexes (ES complexes) formed and so the enzyme activity will be zero.
- As the substrate concentration increases, more and more ES complexes are formed per second and so the enzyme activity increases.
- When the substrate concentration is so high that all the active sites are occupied at any given moment, the reaction cannot proceed any faster and the enzyme activity remains constant at its maximum rate for these conditions.

This explanation takes account of all the changes in activity that are evident in the graph and explains them in a way that is reasonable to expect from an AS biology student. The following are important when explaining the trends in results:

- Make sure you have the trends clearly in your mind.

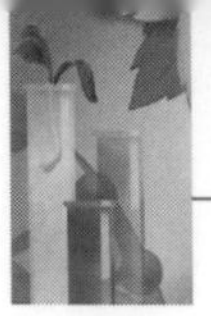

- Think clearly about what is happening when the DV changes. You may have to 'unoperationalise' the DV to do this — for example, increasing amounts of oxygen being collected (per unit of time) as catalase catalyses the decomposition of hydrogen peroxide translate into increasing activity of catalase.
- Try to think why a change in the IV should change the unoperationalised DV.

When drawing conclusions be careful not to suggest a causal relationship when one has not been proved. Ecological investigations often show a correlation between two variables, but, because other variables are not controlled, cause and effect is not proven. Consider the data in Table 20 about leaves from nettle plants growing in the light and others growing in the shade.

Table 20

	Surface area/cm^3	
Condition	**Total of 30 leaves**	**Mean**
Shade	796	26.53
Light	680	22.67

There seems to be a negative correlation between light intensity and surface area of nettle leaves. However, there are other uncontrolled factors, so we are not justified in saying that change in light intensity causes a change in leaf size. We must just report the negative correlation and leave it there.

Evaluating investigations

When scientists report their investigations to other scientists (usually in scientific journals), they have to make clear the limitations of their investigations. They should include an evaluation of:

- the limitations of their equipment
- the limitations of their procedures
- the reliability of their results
- the validity of their conclusions

Evaluating equipment

When you come to evaluate the equipment you are using, think about the following. We can use the collection of oxygen from the decomposition of hydrogen peroxide by catalase as an example (see Figure 10).

Ask yourself the following questions:

(1) What is the equipment supposed to do in my investigation *if it works perfectly*?
Answer: 'if it works perfectly' is an important caveat. If the gas syringe connected to the reaction vessel in Figure 10 works perfectly, it will collect all the oxygen released by the reaction.

(2) How might it not be working perfectly?
Answer: there might be leaks or delays that would result in some of the oxygen escaping.

(3) What impact might this have on my results?
Answer: this would result in all the readings being lower than they should be.

Bear in mind that graduated (measuring) equipment is not perfect. As a rough guide, whatever the graduations are, there could be an error of half of one of these graduations. For example, if you measure 5 cm^3 in a syringe marked in 1 cm^3 graduations, the volume could be anything between 4.5 cm^3 and 5.5 cm^3. As a percentage of the volume being measured, this is quite an error. 0.5 cm^3 as a percentage of 5 cm^3 is 10%. So, there could be a built-in error of 10% in all your results, simply because of the equipment you chose to use.

There are two solutions to this particular situation:

- Use equipment capable of greater precision (say a syringe or pipette with 0.1 cm^3 graduations).
- Use larger volumes — a 0.5 cm^3 error as a percentage of 20 cm^3 (rather than 5 cm^3) is an error of only 2.5% (rather than 10%).

You may be able to calculate the percentage error for several different pieces of equipment used in the investigation. Do not try to give an overall percentage error by adding these together. A simple addition is not appropriate — the overall assessment is much more complicated. We cannot go into the maths of this here, so trust me. Report the individual percentage errors and leave it at that!

Evaluating procedures

When evaluating procedures, you should use a similar logic to evaluating equipment. Again, we can use the catalase investigation as an example. You should ask yourself the following questions:

(1) What is the purpose of each procedure, if I carry it out perfectly?
Answer: in the catalase investigation, using a stopwatch allows you to find how long it takes to collect a certain volume of oxygen; you must start the stopwatch as soon as the hydrogen peroxide is added to the catalase.

(2) Why might I not be able to carry it out perfectly?
Answer: turning the tap on the burette and starting the stopwatch as soon as the hydrogen peroxide mixes with the catalase may not be physically possible. You will also have to judge when the volume in the gas cylinder has reached (say) 30 cm^3.

(3) How might this affect my results?
Answer: depending on how you choose to solve the problems above (and you must be consistent in this — it must be a standardised procedure), you could record a time that is too short (if you start the stopwatch fractionally after the mixing has started), or too long (if you start the stopwatch and then turn the burette tap).

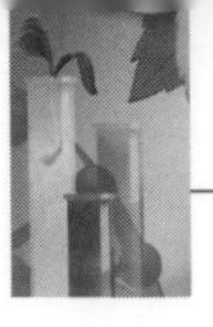

Tip You must be careful to avoid saying something like: 'I could have measured the mass more carefully'. This is not an evaluation — it is an admission of carelessness. If you *could* have measured more carefully, you *should* have. End of story.

Evaluating data

There are several questions you can ask that influence the reliability of the data. They vary slightly according to the type of investigation. Points to consider for laboratory investigations are shown in Table 21 and points to consider for ecological investigations are shown in Table 22.

Table 21

Point to consider	Possible consequences
Did I use enough data points across the range (e.g. did I investigate enough temperatures)?	Too few means that the individual data points will be more widely spread and interpolation (assuming the trend between two points) will be less reliable.
Were the data points appropriately spaced (e.g. did I space the temperatures every 10°C)?	Inappropriate spacing of the data points affects interpolation. Where there are wide gaps, interpolation of data is much less reliable.
What do the standard deviations show?	If you have calculated standard deviations, then large values (as a proportion of the mean) indicate a lot of variability in your data — they are less reliable than data with smaller standard deviations.
Did I carry out enough repeats?	Too few repeats means that the mean will be unduly affected by anomalous data; the risk of the data being unrepresentative is increased.
Did I measure to an appropriate degree of accuracy?	If the measurements were less precise than they could have been, there will be a larger percentage error in the data, decreasing the reliability of the data.
Did I have to use personal judgement at any point?	This is sometimes unavoidable (e.g. if you have to judge when a colour is formed or disappears). It is a limitation and *will* introduce a degree of variability in your data. Include in your evaluation any steps you took to minimise the effect (e.g. using colour standards).
Did I standardise all my procedures?	Even if you did, there may still be an error in your results. (If, in the catalase investigation, you start the stopwatch fractionally after the mixing has started, you will record a time that is too short.) This is a **systematic error** — it will affect all results in a similar way and you can allow for this. Varying the procedures introduces *unsystematic* errors that you cannot allow for and which make your results less reliable.
Did I control all the variables I could have controlled?	This is more likely to affect the validity of the investigation than the reliability. You may still produce repeatable (reliable) results, but, if you did not control (say) the pH, then the results will not be a measure of what you intended, so they will not be valid.

Table 22

Point to consider	Possible consequences
Was the overall sample size large enough (Did I take enough readings?)	A small sample is less likely to be representative of the population as a whole and will be more easily affected by anomalous data.
For an investigation into numbers in an area: • Were the sample points randomly allocated? • Were there sufficient sampling points?	• Non-random allocation of the sampling points introduces bias into the investigation. Bias reduces both reliability and validity because some areas/individuals are more likely to be included than others. • Too few points means anomalous results will have a large effect on the mean and, therefore, on the estimate of overall numbers.
For an investigation into distribution across an area: • Were the sampling points allocated systematically (e.g. every 10 metres)? • Were there sufficient sampling points?	• Systematic allocation of sampling points (systematic sampling) is the best way to observe changes in abundance across an area; random allocation of sampling points could miss a key section (see Figure 23). • Too few points means that some species and some changes in abundance could be missed.
What do the standard deviations suggest?	Large standard deviations (as a proportion of the mean) suggest considerable variability in the data and, therefore, less reliability.
Did I control the variables that could be controlled?	If you measure the light intensity of two areas that could be in sunlight or shade at different times of the day, did you measure them both in sunlight (or shade)? Failing to do this will introduce serious non-systematic errors and reduce the validity of your conclusions.
Did I monitor the main variables that I could not control?	It is impossible to control all variables in the field. However, several factors may (and probably will) vary between the sampling points. You should monitor as many of these as possible so that you can take into account the possible impact of changes in these variables when drawing your conclusions. Not monitoring key variables means that there could be effects of which you are unaware and this would reduce the validity of your conclusions.

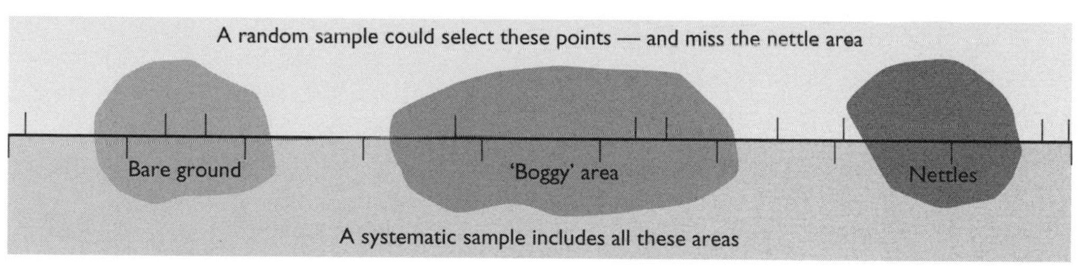

Figure 23 Systematic and random sampling along a transect

An evaluation must *not* be entirely negative. State the positive things also. For example, if you *did* measure to an appropriate degree of accuracy, then say so (as long as you can justify it). Then, state the impact of these on the reliability and/or validity of your investigation.

What might a Unit 3 ISA (or EMPA) task look like?

AS Biology ISA (EMPA) Biol3 task sheet

The effect of substrate concentration on the activity of an enzyme

Introduction

The activity of enzymes can be measured in several ways. Catalase is an important enzyme in living cells as it breaks down hydrogen peroxide into water and oxygen. In this task you will investigate the changes in the amount of oxygen given off when catalase is mixed with hydrogen peroxide solutions of different concentrations.

Outline method

You are provided with:

- five hydrogen peroxide solutions, each with a different concentration
- catalase solution
- conical flask
- burette/rubber bung assembly for the conical flask
- trough
- 100 cm^3 measuring cylinder
- stopwatch
- two 20 cm^3 syringes

1 Place 20 cm^3 of the least concentrated solution of hydrogen peroxide in a conical flask.
2 Assemble the apparatus as shown in Figure 24.

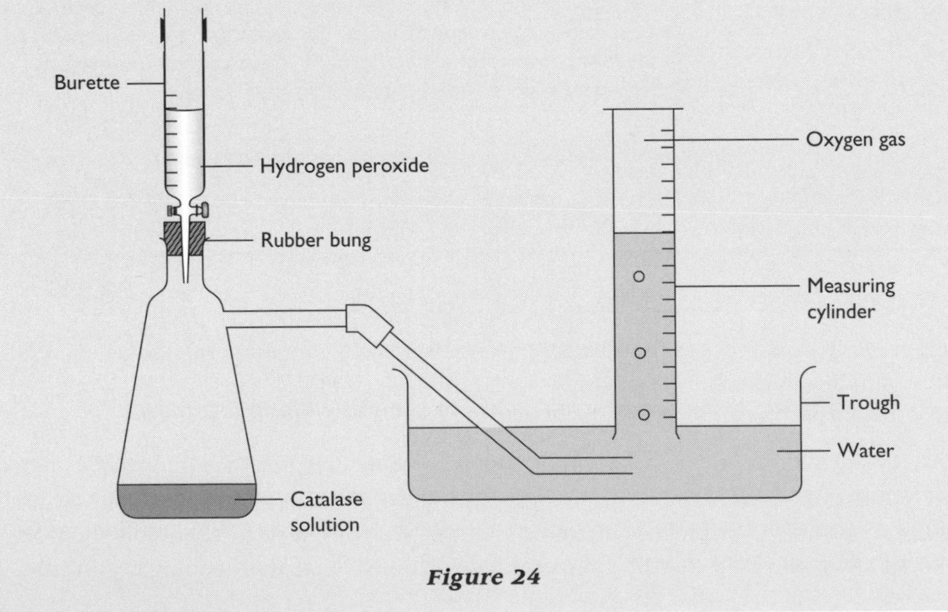

Figure 24

3 Make sure the measuring cylinder is full of water.
4 Add catalase to the hydrogen peroxide.
5 Repeat this procedure for all the concentrations of hydrogen peroxide.
6 In this investigation you should decide for yourself:
   - what data to collect in order to allow you to calculate a rate of activity
   - how many repeats to use with each concentration of hydrogen peroxide
   - how to control variables that might influence the data to be collected

Use the space below to process your data.

Use the graph paper to plot a graph of your processed data.

Hand in this sheet at the end of the practical session.

You have to:
- carry out the investigation
- record your results in a suitable table
- process your data and plot a suitable graph

Carry out the investigation

The basic instructions are given. In order to calculate a rate of activity, you must measure both the volume of oxygen collected and the time taken. One of these must be constant (a controlled variable) for each repeat — it does not matter which. The person marking your work will be able to tell from your table whether or not you have done this.

You need to control appropriate variables. Table 4 states that enzyme-controlled reactions are influenced by:
- temperature
- pH
- substrate concentration
- extent (time) of reaction

Here, you are investigating substrate concentration, so you should try to control the other variables. You will probably be asked about this in the ISA (or EMPA) written paper.

Record your results in a suitable table

Remember that the layout of a table (see Table 10) should have the IV in the first column and results for repeats of the DV in other columns; units should be in the heading and not in the body of the table. You could use a table similar to Table 23.

Notice that a column for 'mean' is included. You will not need this initially, as the mean is part of your processed data. However, you will need it when you come to work out rates ready to plot your graph. Notice also that the heading for the DV

results takes account of both the volume of oxygen collected and the time taken — in this case, how much oxygen is collected in a set time (time is a controlled variable). However, it could have been time taken to collect a certain volume of oxygen. As the experiment is to measure volume in a set time, the units are volume units — cm^3.

Table 23

Concentration of hydrogen peroxide/%	Volume of oxygen collected in 60 seconds/cm^3			
	Repeat 1	**Repeat 2**	**Repeat 3**	**Mean**
Concentration 1				
Concentration 2				
Concentration 3				
Concentration 4				
Concentration 5				

Process your data and plot a suitable graph

Your must always plot the IV on the *x*-axis against the DV on the *y*-axis. In this investigation:

- IV is the concentration of hydrogen peroxide
- DV is the rate of reaction of catalase

To present data as rates, they must be in the form of 'something per unit of time' — for example, miles per hour or kilometres per second are both rates of movement. Here, it is 'volume of oxygen given off per second' (or per minute if you prefer). To get the data in this form, you need to:

- calculate the mean for each concentration
- divide this mean by 60 to give a mean volume per second

You can now plot your graph. An example is given in Figure 25.

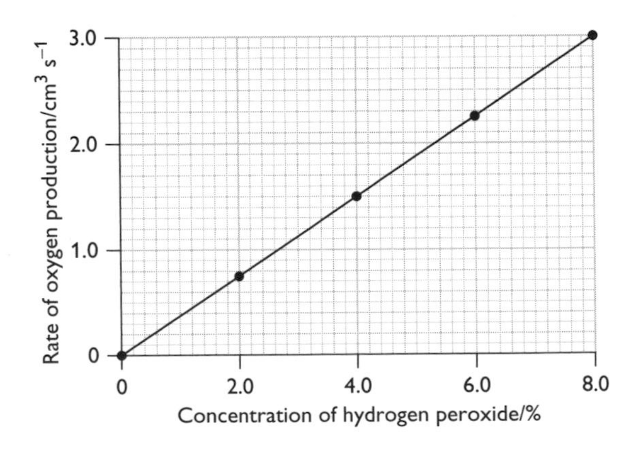

Figure 25

The written task

In Section A of the ISA (or EMPA) written test, you have to answer questions about your investigation, or results based on a similar investigation.

Section A

These questions are about your investigation on the effect of substrate concentration on the activity of an enzyme.

You should use the task sheet, your results and the statistical calculations you have carried out to answer these questions.

Answer *all* questions in the space provided.

1 What was the independent variable (IV) in your investigation? (1 mark)

e The independent variable in this investigation is the concentration of hydrogen peroxide.

2 Describe how you controlled *two* other variables in your investigation. (2 marks)

e You could here describe techniques such as the use of a buffer to control pH, or a water bath to control temperature.

3 You were told to make sure that the measuring cylinder was full of water. Explain the importance of this. (2 marks)

e If there had been any air in the measuring cylinder, then the measurements of the volume of oxygen could be inaccurate. It is important to go on to say that this would influence the reliability of the results.

4 Explain two benefits of carrying out repeat investigations. (3 marks)

e The main benefits are that repeats allow the identification of anomalous results and make it possible to take an average value, which is likely to be more representative than any single value.

5 A student carried out a similar investigation to you. Figure 26 shows how this student plotted her results.

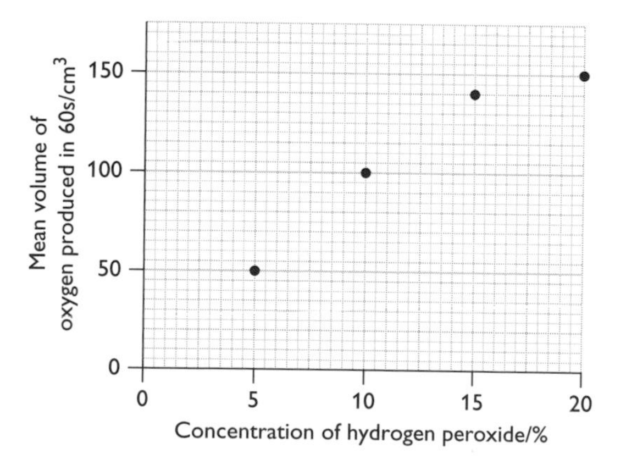

Figure 26

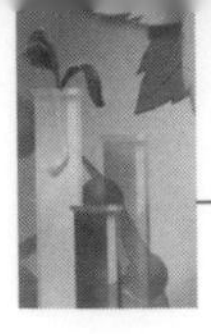

Tip A second graph is supplied to make sure that everyone is on a level playing field for this part of the assessment. Mistakes can occur when carrying out investigations and it is not always possible to rectify them. This ensures that mistakes are not carried forward into the next stage of the ISA (or EMPA).

(a) Use the points to plot a curve. (2 marks)

It is important to remember that you can either:

- join all the plots with ruled lines, *or*
- where there is a clear trend, draw a curve of best fit

In this situation, either would be acceptable. Remember that there are no data for 0%, so you must not extrapolate your curve to zero.

(b) The units chosen for the y-axis by the student are not entirely suitable. Explain why. (1 mark)

The investigation is about the activity of an enzyme, which is normally described in terms of a rate of production of some product. A rate is normally expressed per unit time — so much in 1 minute, 1 second, 1 hour. The unit for the *y*-axis is volume collected in 60 seconds, which is 1 minute, but is not expressed in that way. It's not exactly wrong, but it could be better.

(c) Use your curve to find the volume of oxygen produced when the concentration of hydrogen peroxide solution was 7%. (1 mark)

In this situation the examiner will try not to 'carry forward an error'. If you have made an error in drawing the curve, but read information correctly from your 'wrong curve', you will be credited with full marks.

(d) Describe, and use your biological knowledge to explain, the shape of the curve between the following concentrations of hydrogen peroxide:
(i) 5% and 15% (2 marks)
(ii) 15% and 20% (2 marks)

Notice that you have to do two things for each part of the question — you must *describe* the curve and *explain* it. Between 5% and 15%, there is a linear (regular) increase in the rate as more and more active sites become occupied at any one time. Between 15% and 20%, there is very little increase because all the active sites are occupied for most of the time.

(e) Conclusions drawn from this graph may be limited and not very reliable. Suggest why. (2 marks)

Once you see a question about reliability, you should be thinking about lack of repeats or inappropriate values for the IV. There is nothing that specifically indicates a lack of repeats in this investigation and since the *y*-axis is labelled 'mean volume of oxygen produced', some repeats were carried out. However, there are only four quite widely spaced values of the IV and there are no data for 0% (or for any value less than 5%) so what happens here is a complete guess.

Section A total: 18 marks

Section B

In section B of the written test, you will be given resource materials related to the investigation you have carried out. These are often of an applied nature — that is, the concepts are in a real-life situation. For example, you could be given resource material about the activity of catalase in liver cells from humans.

Resource sheet

Resource A

Biologists investigated the relative catalase content of three cheeses:

- Swiss Emmental cheese
- Raw Cheddar cheese
- Pasteurised Cheddar cheese

Their procedure is described below:

- Grind 5 grams of cheese (freshly cut from the interior of the cheese) with 25 grams of sand.
- Place this material in a large test tube.
- Add 25 cm^3 of 1% hydrogen peroxide buffered to pH 7.
- Place the tube in a water bath at 20°C.
- Collect the oxygen liberated in a water-filled burette.
- Measure the volume released after 30 minutes.

Their results are shown in the table.

Sample number	Volume of oxygen released in 30 minutes/cm^3		
	Unpasteurised Cheddar cheese	**Pasteurised Cheddar cheese**	**Swiss Emmental cheese**
1	5.0	0.5	10.5
2	6.5	0.5	37.5
3	7.5	1.5	12.5
4	5.0	0.5	17.0
5	8.5	0.5	11.0
6	4.0	0.6	30.5
7	3.5	0.8	5.0
8	2.0	0.5	21.0
9	5.0		24.0
10	8.0		17.5
11	3.0		19.0
12	6.0		18.0
13	4.5		28.5
14	6.0		8.0
Mean		0.675	18.6

(Contd)

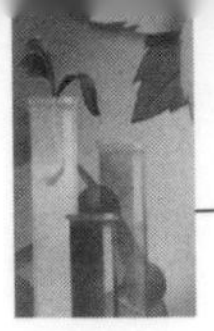

Resource B

Pasteurisation is a process that involves heat treatment and kills many harmful bacteria. Milk used in the production of cheeses is often pasteurised. However, in the production of Swiss Emmental cheese, the milk is treated with hydrogen peroxide instead of by pasteurisation. Hydrogen peroxide is toxic to cells.

However, residues of hydrogen peroxide in the milk would inhibit the bacterial cultures that are required for the cheese production, so all traces of it must be removed. Catalase enzymes are added to convert the hydrogen peroxide to water and oxygen. This catalase was obtained traditionally from cattle livers.

A second group of biologists investigated catalase production by bacteria used to make Emmental cheese in modern production techniques. Their results are shown in the table below.

Type of bacterium	Volume of oxygen produced in 30 minutes/cm^3
Type 1	0.0
Type 2	0.0
Type 3	0.0
Type 4	0.0
Type 5	48.7
Type 6	0.0
Type 7	0.0

Use the information in the **Resource sheet** to answer the questions.

Answer *all* questions in the space provided.

Use **Resource A** to answer question 6.

6 (a) Calculate the mean volume of oxygen produced by unpasteurised Cheddar cheese. (1 mark)

A simple starter — just add up the volumes and divide by 14.

(b) List four variables that were controlled in the investigation. (2 marks)

You might think that this is 1 mark for two variables, but it is more likely to be marked as follows: four correct = 2 marks, two or three correct = 1 mark, none or one correct = 0 marks. There are a number of controlled variables you could choose:

- mass of cheese
- mass of sand

- volume of hydrogen peroxide
- pH
- temperature
- time
- source of cheese (from the centre)

This should be an easy 2 marks.

(c) Compare the reliability of the data for Swiss Emmental cheese and raw Cheddar cheese. (3 marks)

The obvious points to make here are that the data for the Emmental cheese are quite varied. There is a considerable range between highest and lowest, whereas results for the raw Cheddar are closer together. However, you should consider not just the actual differences — bigger numbers are likely to have bigger differences — but also the proportions of one to the other. For example, the ratio of highest value to lowest value in the Cheddar cheese data is 8.5:2.0 = 4.25. That's quite a high ratio. However, the ratio for the Emmental is even higher at 7.5 (37.5:5.0). In this case, looking at these ratios supports what we found with the raw data. In other cases it might not and you should point that out.

(d) The mean values for the three cheeses look very different.

(i) What other information would you require to be sure that there is a real difference in these means? (1 mark)

You would also need to know the standard deviations.

(ii) How would you use this information to decide whether or not the differences in the means were real? (2 marks)

You would need to know whether or not the standard deviations overlap. If they do, then the differences are unlikely to be real. If they do not overlap, the differences are likely to be real.

Use **Resource B** to answer question 7.

7 (a) Suggest why hydrogen peroxide can be used instead of pasteurisation in the production of cheese. (2 marks)

You are told that hydrogen peroxide is toxic to cells. Presumably this means all cells, including those of any harmful bacteria in the milk. The hydrogen peroxide will kill these bacteria and sterilise the milk.

(b) Some of the flavour of Emmental cheese comes from the proteins in the milk used. Suggest why producers of Emmental prefer to use hydrogen peroxide rather than pasteurisation in the production of their cheese. (3 marks)

Pasteurisation involves heating the milk. Heat denatures proteins and so pasteurisation could affect the flavour of the cheese. Using hydrogen peroxide will not affect the proteins and so will not affect the flavour.

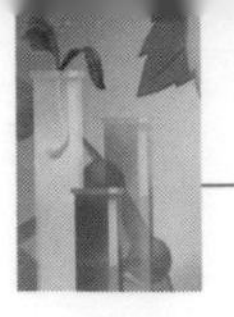

(c) Explain why catalase is added to Emmental cheese. (3 marks)

Catalase decomposes the hydrogen peroxide into water and oxygen. If it were not added, the hydrogen peroxide would kill the bacteria that are added to the milk in the production of the cheese.

(d) Suggest why the producers of Emmental no longer need to use catalase from the livers of cattle. (1 mark)

One type of the bacteria that they use in the production of the cheese produces catalase.

Section B total: 18 marks

The Unit 6 assessment

What you must be able to do

This unit addresses the following aspects of the A2 subject criteria. The ability to:

- demonstrate and describe ethical, safe and skilful practical techniques, selecting appropriate qualitative and quantitative methods
- make, record and communicate reliable and valid observations and measurements with appropriate precision and accuracy
- analyse, interpret, explain and evaluate the methodology, results and impact of your own and others' experimental and investigative activities in a variety of ways

Candidates will be assessed on their understanding of investigative and practical skills in this unit and in Units 4 and 5. Opportunities to carry out practical work are provided in the context of material contained in Units 4 and 5.

3.6.1 Investigating biological problems involves changing a specific factor, the independent variable, and measuring the changes in the dependent variable that result.

The requirements here are identical to those for the Unit 3 assessment except for the following:

- Describe how you would collect a full range of useful quantitative data that could be analysed statistically.

3.6.2 Implementing involves the ability to work methodically and safely, demonstrating competence in the required manipulative skills and efficiency in managing time. Raw data should be collected methodically and recorded during the course of the investigation.

The requirements here are identical to those for the Unit 3 assessment.

3.6.3 Data should be analysed by means of an appropriate statistical test. This allows calculation of the probability of an event being due to chance. Appropriate conclusions should be drawn and scientific knowledge from the A-level specification should be used to explain these conclusions.

This means that you should be able to use your knowledge and understanding of the A2 specification in order to:

- select an appropriate statistical test from the following:
  - standard error and 95% confidence limits
  - Spearman rank correlation
  - χ^2 test
- justify your choice of the statistical test
- construct an appropriate null hypothesis

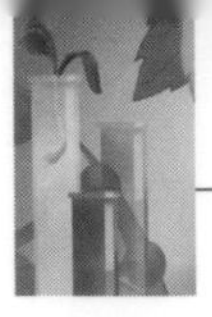

- calculate the test statistic given a standard scientific calculator
- interpret the calculated test statistic in terms of:
  - the critical value at the 5% significance level ($p = 0.05$)
  - chance and probability
  - acceptance or rejection of the null hypothesis
- draw valid conclusions, and use your biological knowledge and understanding of the A-level specification to explain and justify these conclusions

Tip Notice that the last statement refers to the A-level specification. This means that you may need to include material from the AS units as well as from the A2 units.

3.6.4 Limitations are inherent in the material and apparatus used and procedures adopted. These limitations should be identified, evaluated and methods of overcoming them suggested.

Candidates should be able to:

- identify the limitations inherent in the apparatus and techniques used
- discuss and assess the relative effects of these limitations on the reliability and precision of the data and on the conclusions that may be drawn, resolving conflicting evidence
- suggest realistic ways in which the effect of these limitations may be reduced
- suggest further investigations that would provide additional evidence for the conclusions drawn

Unit 6 assessment: ISAs and EMPAs

There are two routes through which Unit 6 assessment can be delivered. They are:

- **ISA (Investigative Skills Assessment)** Your teachers mark your practical assessments and submit the marks, and a selection of work as examples, to a moderator who checks the marking.
- **EMPA (Externally Marked Practical Assignment)** All the practical work you complete is marked by examiners; none of it is marked by your teachers.

The components of the two assessments and their approximate mark allocations are shown in Table 24.

Table 24

Assessment route			
ISA		**EMPA**	
Component	**Marks**	**Component**	**Marks**
Practical skills assessment: your teacher submits an assessment based on your practical skills throughout the course.	6	Practical skills verification: your teacher submits a verification of your practical skills, based on your work throughout the course.	0

Assessment route			
ISA		EMPA	
Component	Marks	Component	Marks
Task Stage 1: you carry out a task specified by AQA and record your results in a table. You have to decide on some aspects of the task yourself (e.g. controls, volumes, number of repeats). There are no marks for this part of the assessment at A2.	0	Task 1: you carry out a task specified by AQA. You have to decide on some aspects of the task yourself (e.g. controls, volumes, number of repeats). You answer questions about some aspects of the task.	10
Task Stage 2: you begin an analysis of your data and carry out an appropriate statistical test. AQA supply guidance to help you: • choose the test • carry out the calculation	6	Task 2: you carry out a second, related task specified by AQA and record the results in a table that you construct. (4 marks) You begin an analysis of your data and carry out an appropriate statistical test. AQA supply guidance to help you: • choose the test • carry out the calculation (6 marks)	10
Section A of written test: you answer questions based on the results from your own investigation.	18	Section A of written test: you answer questions based on the results from your own investigation.	15
Section B of written test: you answer questions based on related material supplied by AQA.	20	Section B of written test: you answer questions based on related material supplied by AQA.	15
Total for ISA	50	Total for EMPA	50

The practical skills you will need

AQA assumes that you are familiar with the use of basic equipment, such as measuring cylinders, Bunsen burners, thermometers etc. as well as the skills developed in the AS course. But you must be able to show the following skills also.

The use of a three-way tap in collecting gas samples

The apparatus shown in Figure 27 is a respirometer. It is used for measuring the rate of oxygen uptake by small, respiring organisms. One of the tubes contains the respiring organisms; the other acts as a 'thermobarometer'. Any changes in pressure and temperature that take place in the experimental tube (which would affect measurements of volumes), also take place in the thermobarometer tube and so are effectively cancelled out. The bung of each boiling tube is fitted with a three-way tap. This makes it possible to take repeat readings without having to reassemble the apparatus, and it also eliminates many of the leaks that could otherwise occur.

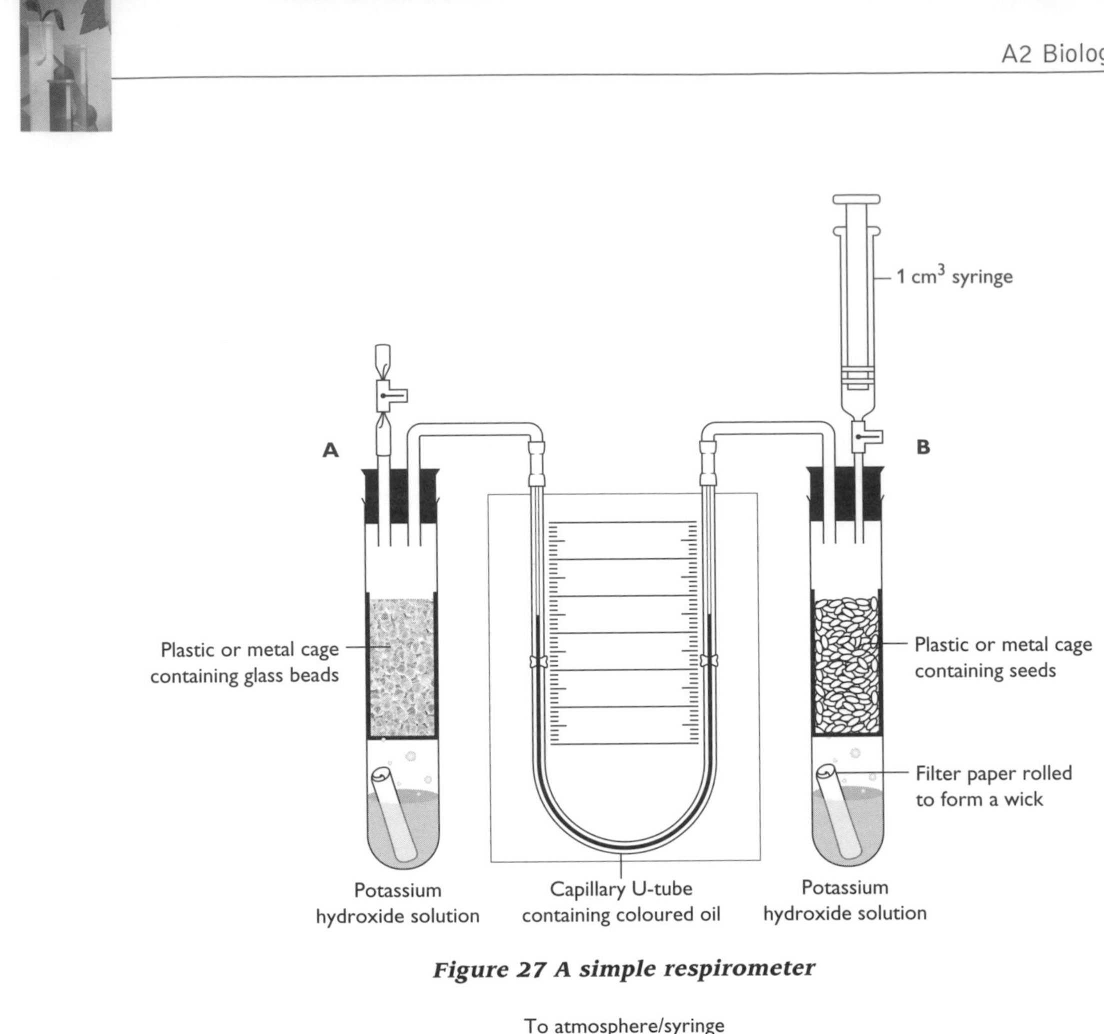

Figure 27 A simple respirometer

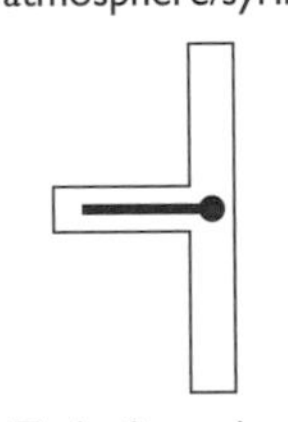

Figure 28 How a three-way tap works

When the tap is in the position shown in Figure 28, the atmosphere and the boiling tube are connected. The position of the tap can be altered so that any two components are connected. By opening the three-way tap on tube A to the atmosphere and that on tube B to the syringe, and moving the barrel of the syringe, the pressure in the two tubes can be equalised.

Establishing anaerobic conditions

In experiments involving fermentation (usually by yeast), it is important that anaerobic conditions are maintained throughout the investigation. The reason for this is that the rate of fermentation is usually measured by recording the rate of carbon dioxide production under different conditions. This is valid only if fermentation is the sole method of production of carbon dioxide.

The summary equation for fermentation in yeast is:

$$C_6H_{12}O_6 \rightarrow 2C_2H_5OH + 2CO_2$$

However, the summary equation for aerobic respiration in yeast is:

$$C_6H_{12}O_6 + 6O_2 \rightarrow 6H_2O + 6CO_2$$

Allowing aerobic respiration to occur introduces several potential errors. For example:

- Aerobic respiration gives three times as much carbon dioxide per molecule of glucose; so, if different amounts of aerobic respiration took place in different trials, the results would not be valid.
- Collecting the carbon dioxide produced often depends on the pressure of the carbon dioxide being able to force the gas through a delivery tube; with aerobic respiration, an equal volume of oxygen is absorbed by the yeast, so there is no overall change in pressure.

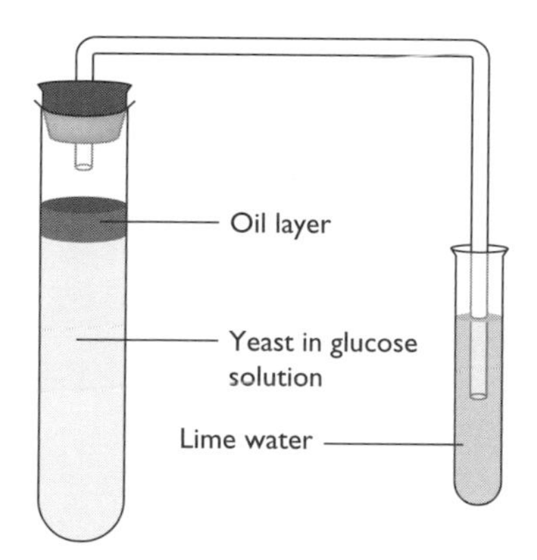

Figure 29 Anaerobic respiration in yeast

Figure 29 shows a simple way to investigate the rate of anaerobic respiration in yeast. To maintain anaerobic conditions, you must take two main precautions:

- Make up the glucose solution in boiled, cooled, distilled water (boiling removes dissolved oxygen).
- Once in the reaction vessel, with the yeast added, cover the surface of the liquid with a thin layer of oil or liquid paraffin (this prevents oxygen in the atmosphere from entering the liquid).

Carbon dioxide can still escape through the oil/liquid paraffin because it is at higher pressure than the atmosphere.

Random sampling

Sampling populations

Investigating all the members of a population is only rarely possible. So to form a picture of what that population is like, we need a **representative sample** — a sample that is typical of the whole population. A sample is a subset of a population.

How do we take a representative sample? This involves a circuitous argument because to take a truly representative sample we would have to know already what the population was like — for example, what percentage of the main types of individuals there were in the population, in order to include the same percentages in the sample. However, since we do not know that we must settle for an approximation. The best we can do is to take a random sample. In a **random sample** every individual has an equal chance of being included in the sample, so there is no **bias**.

Tip Bias in investigations results in a sample that systematically misrepresents the nature of the population. For example, if a biologist selects an area of buttercups to be the sample because 'they look typical', bias is introduced. The perception of what is typical may be misinformed and so the sample will be biased.

However, you must appreciate that a random sample does not *guarantee* a representative sample. Every individual has an equal probability of being included in the sample, so the sample *could* be made up of all the tallest individuals, or all the heaviest mice. It is unlikely, but it could happen. In any case, it's our best shot at a representative sample.

Suppose you want to find the average height (or mass, or IQ, or handspan) of 400 17-year-olds. To create a random sample:

- Give each 17-year-old a number.
- Use a random number generator on a calculator or computer to generate a sample of (say) 50 numbers between 1 and 400.
- Measure each individual in the sample and work out the mean.

The same principle could apply to any population.

To improve reliability, you could:

- take three random samples, make measurements of the desired feature, calculate the mean and calculate the mean of the means
- take one larger sample

Sampling areas

To estimate the abundance of each species in an area:

- Estimate the abundance in several sample areas.
- Calculate the mean abundance.
- Multiply this by the ratio of the whole area to that of the sample area.

The abundance of organisms in an area is usually estimated by using quadrat frames. The quadrat forms the sample area. Quadrat frames come in different styles and sizes. The simplest is a square metal frame, which you place on the ground and estimate the abundance (see p. 65) of the organisms inside the quadrat frames. Figure 30 shows how to place quadrat frames at random.

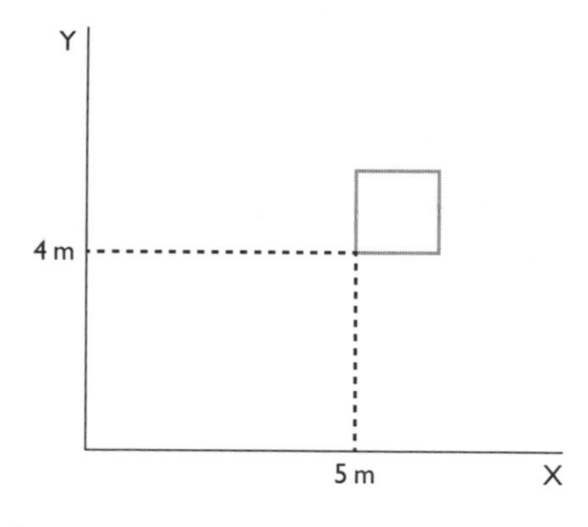

Figure 30 Placing a quadrat at random

A single number is not enough to define the position of the quadrat — we need two. What we must do is:

- Divide the area into a grid.
- Use the random number generator on a calculator to produce a pair of coordinates, for example X5, Y4 (this pair would define the bottom left-hand corner of the square on the diagram).
- Place a quadrat with its bottom left-hand corner on the intersection of the coordinates.
- Estimate the abundance of the different organisms inside the quadrat.
- Repeat several times and calculate the mean.

Collecting reliable ecological data

Ecological investigations are usually aimed at finding out something along the following lines:

- What types of organisms are found here?
- How abundant are they?
- Where are they most abundant within the area?
- What abiotic (environmental) factors might be responsible for the distribution?

To find the answers to these questions, some or all of the following must be measured or estimated:

- size of populations
- distribution of organisms
- values for the abiotic factors

Estimating population density (the size of a population)

A method for plants and small non-mobile animals

- Place quadrats randomly as described above.
- Count the number of organisms of the species you are investigating in each sample quadrat.
- Find the mean number per quadrat.
- Estimate the population size using the formula:

$$\frac{\text{mean number of organisms per quadrat} \times \text{area of field}}{\text{area per quadrat}}$$

It is important that you use the same units for the area of the field and the area of the quadrat (e.g. both in cm^2 or both in m^2).

A method for small mobile animals

To estimate the size of a population of animals that move about, biologists use the **mark-release-recapture** technique. It is carried out as follows:

- Collect a sample of the animals from the area and count them (N_1).
- Put a small mark using a harmless substance in an unobtrusive place on each animal. This is to ensure that the animal is not harmed by the process itself and that the marking does not affect its survival chances by making it more noticeable to predators.
- Release the marked animals and allow time for them to disperse among the population.
- Collect a second sample and note both the total size of the sample (N_2) and the number that are marked (n).

The assumption made now is that the proportion of the population collected in the first sample is the same as the proportion of marked individuals in the second sample. We can write this mathematically, as:

$$\frac{N_1}{X} = \frac{n}{N_2}$$

Rearranging the formula allows calculation of the population size:

$$X = \frac{N_1 \times N_2}{n}$$

Suppose that 50 (N_1) woodlice were caught originally and marked and released, and that later 40 (N_2) were caught of which 10 were marked (n). The estimate of the population size would be:

$$X = \frac{50 \times 40}{10} = 200$$

Assumptions in the mark-release-recapture technique

There are several assumptions in the mark-release-recapture technique, all of which are unlikely to be met fully. They are:

- There are no migrations.
- There is no reproduction.
- There are no deaths.
- Marking does not affect behaviour.
- On release, there is random mixing of the marked and unmarked individuals.
- The second sample is representative of the population as a whole.

Despite not being able to meet these assumptions, the technique gives a reasonable estimate of population sizes.

The reliability of the technique is affected by several factors. We can increase reliability by:

- using larger samples at both stages of the investigation
- allowing sufficient time for the released animals to mix (but not too long as they may then be more likely to migrate or die)
- repeating the investigation several times to obtain a mean

Percentage cover

Sometimes, it is extremely difficult to count the number of organisms, even in the small area of a quadrat. How many blades of grass belong to each plant, for example?

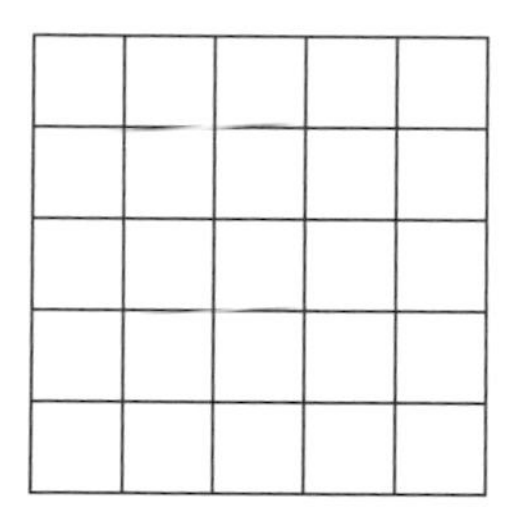

Figure 31

To get around this, biologists sometimes estimate the **percentage cover** of a species. For example, if the organism takes up half of the quadrat, then the percentage cover is 50%. A better way of estimating percentage cover is to use a quadrat that is subdivided into smaller squares.

The quadrat shown in Figure 31 has 25 smaller squares; each square is 4% of the total area.

If the species covers 7 squares, the percentage cover is $7 \times 4\% = 28\%$.

If the species covers 8 squares and 2 part-squares, then a total of 9 squares is a reasonable estimate. This would give $9 \times 4 = 36\%$ cover.

Estimation of frequency

Another method of estimating abundance is to measure the **frequency** of the species. This is the proportion of quadrats in which a species is found, expressed as a percentage. The advantages of using frequency as a measure are that it is easy and fast, and does not involve subjective judgments, which percentage cover can (you must judge how much of the quadrat, or square within a quadrat, is covered).

One major disadvantage of frequency is that results are not only affected by abundance of the species, but also by how it is distributed. For example, two species that are equally abundant will have different frequencies if one species is clumped and the other species is evenly distributed in the community. Figure 32 shows how this can occur.

Species A and species B have an identical population density of seven plants in the $36\,m^2$ area investigated. However, their frequencies are different.

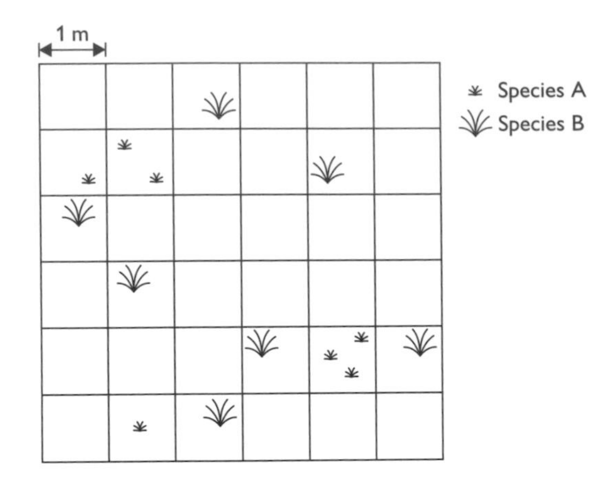

Figure 32 Distribution of a species affects the frequency

Species A occurs in four of the 36 quadrats:

frequency of species A $= \frac{4}{36} \times 100 = 11\%$

Species B occurs in seven of the 36 quadrats:

frequency of species B $= \frac{7}{36} \times 100 = 19\%$

Another problem with using frequency to estimate abundance of a species is that the size of the quadrat used affects the results.

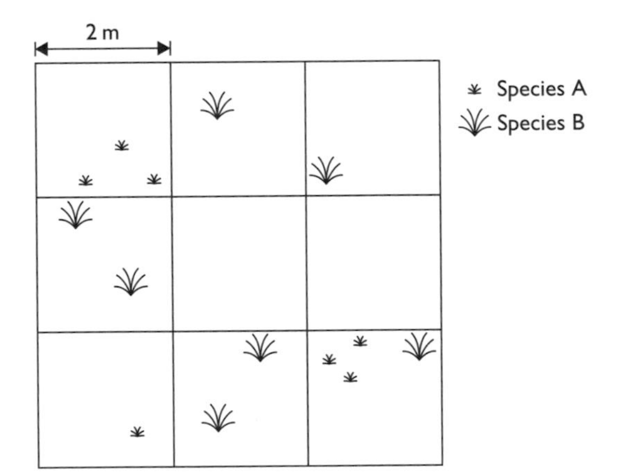

Figure 33 The size of the quadrat influences the estimate of frequency

Figure 33 is similar to Figure 32, except that the quadrats are larger.

Species A occurs in three of the nine quadrats:

frequency of species A $= \frac{3}{9} \times 100 = 33\%$

Species B occurs in five of the nine quadrats:

frequency of species B $= \frac{5}{9} \times 100 = 56\%$

These are very different results from those determined previously, and all that has changed is the size of the quadrat.

Using a line transect to investigate the distribution of organisms in an area

- Lay a tape measure across the sample area.
- At regular intervals (every 4 m in the diagram) lay five quadrats to one side (always the same side) of the tape (Figure 34).

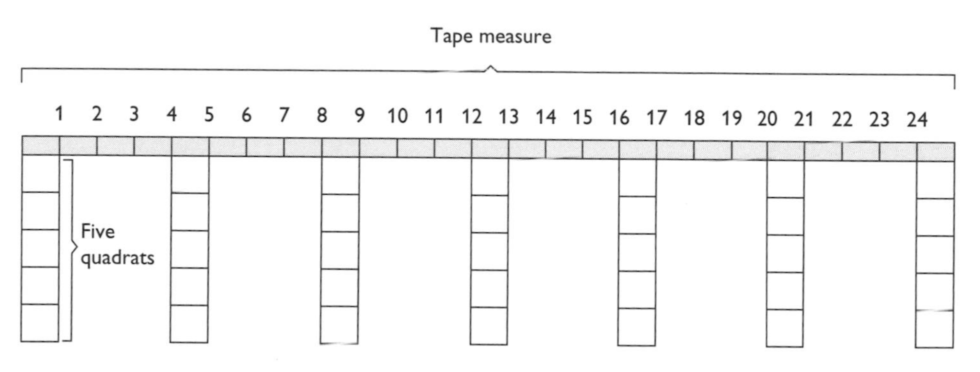

Figure 34 Line transect

- Estimate the abundance of the different organisms at each sampling point along the transect by either:
  - recording presence or absence in each of the five quadrats and converting the number of occurrences to a **percentage frequency** (e.g. a species that occurs in four of the five quadrats has a frequency of 80%)
  - estimating the percentage of each quadrat covered by the species and taking the mean for the five quadrats to give the average **percentage cover**
  - counting the numbers of each organism in each quadrat to obtain a mean for each sampling point

At each sample point, the environmental factors that might influence the distribution of the organisms should be monitored. These might include:

- light intensity
- soil pH
- soil water content
- nutrient availability in the soil

It may be possible to find a correlation between the abundance of an organism and one or more of the abiotic factors using the Spearman rank correlation (see pages 78–81).

The distance between sampling points along the transect is important. If they are too far apart, species will be missed; if they are too close, it will be very time consuming and may yield too many data to be able to see patterns of zonation easily. Table 25 illustrates this point. It shows the number of species recorded for three

different sampling intervals along two 20-metre transects at right angles to each other.

Table 25

Sampling along transect	Number of species recorded	
	Line transect 1 N/S	Line transect 2 E/W
Continuous	31	34
Every metre	15	17
Every other metre	9	12

Measuring abiotic factors

There are always reasons why a particular plant or animal is abundant (or not) in a certain area. These reasons are often linked to abiotic (environmental) factors. For example, some plants are extremely intolerant of an alkaline soil and will not grow well where the underlying rock is limestone (which gives rise to an alkaline soil). Water availability is another key factor for many plants; how many rose bushes grow in the Sahara desert?

There are some ground rules about taking readings of abiotic factors. You should bear in mind that a particular abiotic factor might change over the course of the investigation. Because of the rotation of the Earth, different parts of the area you are investigating could be in shade or in sunlight at different times of the day. Cloud cover might also change. Both these factors could influence light intensity. You need to be alert to this and take all readings in full sunlight (or all in shade, but the former is better). If you take readings of some parts of the area in full sunlight and others in shade (or when there is cloud cover) this particular variable will not have been controlled and your conclusions will not be valid.

If you see a dark cloud on the horizon, you might assume that it will rain before you can complete your investigation. You should collect any soil samples before it rains. Collecting the samples after the shower could have a major effect on your results and make your conclusions invalid.

Specific investigations you need to know about

Besides the practical skills described, there are certain investigations you need to know about. It is unlikely that you will be asked to carry out the particular version of these investigations that you have carried out in your school or college. However, you may be asked to carry out, and comment on, an investigation testing the same relationship in a different context.

Table 26 lists those practicals that are mentioned specifically in the specification.

Table 26

Practical	Independent variable	Dependent variable	Controlled variables
Effect of light intensity on the rate of photosynthesis	Light intensity	Rate of photosynthesis	Carbon dioxide concentration, temperature
Effect of carbon dioxide concentration on the rate of photosynthesis	Carbon dioxide concentration	Rate of photosynthesis	Temperature, light intensity
Effect of temperature on the rate of photosynthesis	Temperature	Rate of photosynthesis	Carbon dioxide concentration, light intensity
The above investigations are examples of the specification requirement that 'candidates should carry out investigations into the effect of a specific variable on an enzyme-controlled reaction'.			
Effect of temperature on the respiration of yeast	Temperature	Rate of respiration	Nature of substrate, substrate concentration, volume of substrate, mass of yeast, aerobic/anaerobic conditions
Effect of nature of substrate on the respiration of yeast	Substrate	Rate of respiration	Temperature, substrate concentration, volume of substrate, mass of yeast, aerobic/anaerobic conditions
Effect of temperature on the respiration of a larger organism (e.g. an insect)	Temperature	Rate of respiration	Mass of organisms, volume of air in respirometer
The above investigations are examples of the requirement that 'candidates should carry out investigations into the effect of a specific variable on the respiration of a suitable organism'.			
Ecological investigation into the effect of an abiotic factor into the numbers of organisms in different areas	Abiotic factor, e.g. soil pH, soil water content, light intensity	Frequency or % cover of the organism in the two areas	Other abiotic factors should be monitored
Ecological investigation into the effect of an abiotic factor on the distribution of organisms along a transect	Abiotic factor, e.g. soil pH, soil water content, light intensity	Frequency or % cover of the organism at points along the transect	Other abiotic factors should be monitored

Other skills you will need

You will need all the 'other skills' described for the Unit 3 assessment. You will also have to be able to select an appropriate statistical test, calculate the test value from your data and interpret your calculated values in terms of probability and chance.

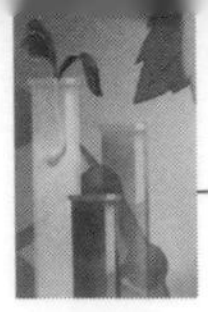

Selecting and using an appropriate statistical test

The statistical test is a means of assessing the probability of patterns in results having a biological cause or just being due to chance events. You will have to:

- choose a statistical test appropriate to the investigation
- calculate the test statistic
- interpret your calculated value in terms of probability and chance

This may sound daunting. However, in the ISA (or EMPA), you will be given guidance on:

- which test to use
- how to calculate the test statistic

To help you choose the correct statistical test AQA provides a statistics sheet, as shown in Figure 35.

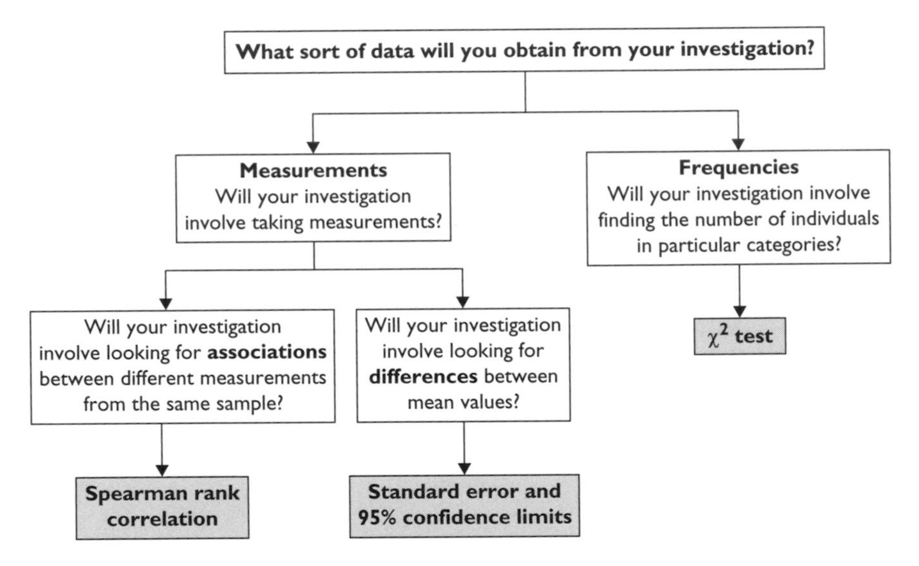

Figure 35 Student's statistics sheet

You need to appreciate:

- the need for a statistical analysis of results
- how to interpret the value of the test statistic once you have calculated it
- that **probability** is a mathematical term which expresses the likelihood of an event occurring; chance describes an event that may occur at random; and that the likelihood of a chance event occurring cannot be calculated

The process of statistical analysis was started in Unit 3 where standard deviations about the mean were used to give an idea of the variability of data. It was noted that if the standard deviations of two mean values of the DV (for different values of the IV) overlap, then we cannot be certain that there is a statistically significant difference between those two means.

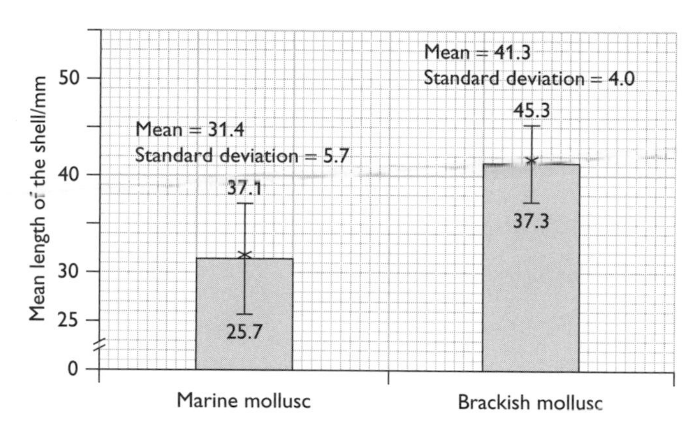

Figure 36 Shell length in molluscs

Figure 36 summarises measurements made on the shell length of two species of mollusc. For each species, the bar chart shows:

- the mean length of the shell
- the standard deviation (shown as bars above and below the mean)

If you look carefully, you will see that the standard deviations do not overlap — it's a close call, but they don't. Therefore, we can be reasonably confident that there is a statistically significant difference between the two mean values. Now look at Figure 37. This bar chart shows the numbers of invertebrates per dm^3 found at three sites in an English lake.

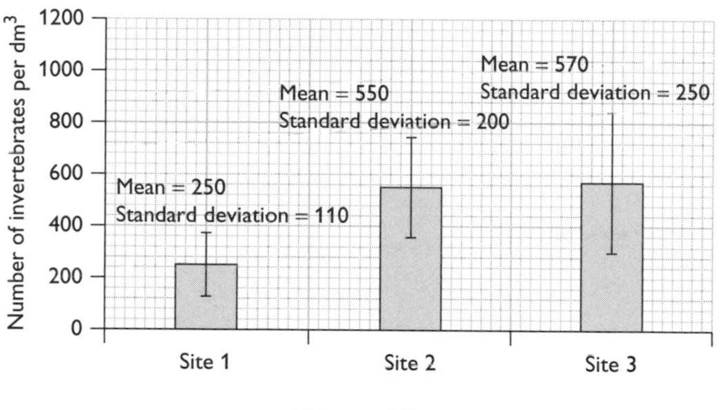

Figure 37

The standard deviations for all three sites overlap, so we cannot be confident that there is a statistically significant difference between the numbers.

What is meant by confident in this context? It is not sufficient to look at the means and say 'I'm pretty confident they are different!' We are talking about statistical confidence; personal opinion is irrelevant.

All statistical tests are designed to test whether or not there is either an association or a statistical difference between two (or more) conditions. We begin from the

point that there is *no* statistical difference between the conditions. This is called a **null hypothesis**. The null hypothesis is the opposite of what you would expect to happen. An **alternate hypothesis** for what you expect to happen can be stated in one of two ways:

- **non-directional** — a change in factor A brings about a change in factor B
- **directional**
  - an increase in factor A brings about an increase in factor B
  - an increase in factor A brings about a decrease in factor B

Table 27 contains some examples of null hypotheses, together with the corresponding alternative hypotheses.

Table 27

	Alternative hypothesis	
Null hypothesis	**Non-directional**	**Directional**
There will be no difference in abundance of buttercups growing in an acid soil and in an alkaline soil.	There will be a difference in abundance between buttercups growing in an acid soil and in an alkaline soil.	There will be more buttercups growing in an acid soil than in an alkaline soil.
A change in temperature will have no effect on the activity of catalase.	A change in temperature will affect the activity of catalase.	An increase in temperature will increase the activity of catalase (up to an optimum temperature).
There will be no difference in size of leaves of nettle plants growing in full sunlight and those growing in shade.	The leaves of nettle plants growing in full sunlight will differ in size from those growing in shade.	The leaves of nettle plants growing in full sunlight will be smaller than those growing in shade.
The type of sugar used will have no effect on the rate of anaerobic respiration in yeast.	Different sugars will be fermented at different rates by yeast.	Glucose will be fermented faster than sucrose by yeast.

Statistics do not deal in certainties, only in probabilities. We can never be certain that a particular null hypothesis should be accepted or rejected. Statistical tests enable us to say, with a certain level of probability, that 'either there are, or there are not'. The level of probability that is commonly used in biology is the 5% probability level, often written as $p = 0.05$. When you calculate a particular test statistic (whatever it may be), there is a certain value of that statistic that corresponds to $p = 0.05$. This is called the critical value and can be found from probability tables. Your calculated value may be greater than, equal to or less than the critical value. You should interpret these outcomes as described below.

- If the calculated value is equal to, or greater than, the critical value for $p = 0.05$, then there is a less than 5% probability that these are just random variations; they are therefore likely to have a biological cause. You reject the null hypothesis.

- If the calculated value is less than the critical value for $p = 0.05$, then there is a greater than 5% probability that the results represent random events. You cannot be confident of a biological cause for any differences; you must accept the null hypothesis.

This is summarised in Table 28.

Table 28

Hypothesis	Test value equal to or greater than critical value	Test value less than critical value
Null hypothesis	Reject	Accept
Alternative hypothesis	Accept	Reject

Statistical tests determine whether, at a certain level of probability, there is no difference. The first example of a statistical test is '**standard error and 95% confidence limits**'. It is calculated using the mean and standard deviation.

Standard error and 95% confidence limits

This test is used to find out whether or not two mean values are significantly different.

Suppose you took a sample of the 17-year-olds in your school and found the mean height. Someone else takes another sample and also finds the mean height. Their sample will almost certainly be different from yours and their mean will be slightly different. So who is right? Well, both of you and neither of you! Without including everyone in the target population, we cannot be certain of the true mean. However, we can estimate the range of values within which the true mean lies. All we need is the mean (already calculated) and the standard deviation, which we can obtain by entering the data into a scientific calculator and pressing the SD button. Standard error is a measure of the variability, not of the raw data, but of the mean itself. It is calculated from the formula:

$$SE = \frac{s}{\sqrt{n}}$$

where

SE = standard error
s = standard deviation
n = sample size/number of measurements

The standard error shows the variability of the mean *not* the variability of the data. It tells us that 68% of the time the mean will fall within the range corresponding to the calculated mean ± the standard error.

Consider an investigation into the reaction times of two groups of individuals. The null hypothesis is:

'There is no difference in reaction time between group 1 and group 2'.

Figure 38 shows the mean and standard error of reactions times in two groups.

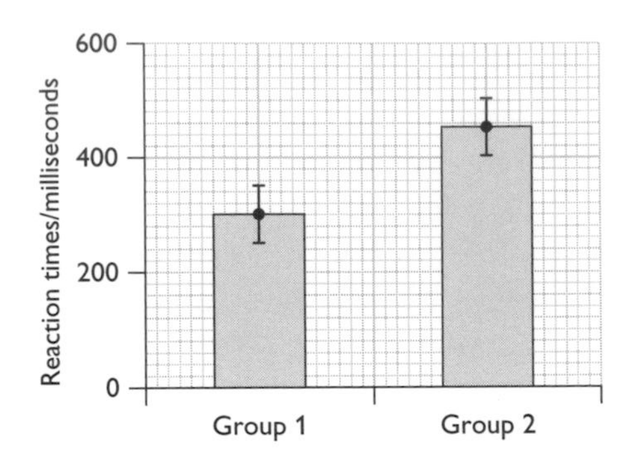

Figure 38 Reactions times in two groups of individuals

This means that 68% of the time, the mean reaction times of these two groups will not overlap because they will fall within the range of values shown on the bar chart. However, our statistic does not mention 68% — it's all about 95% confidence limits. How do we get from 68% to 95%? Easy — we use a range of values equal to 1.96 times the standard error (twice is all right for your ISA or EMPA). Figure 39 shows the same bar chart with the 95% intervals plotted.

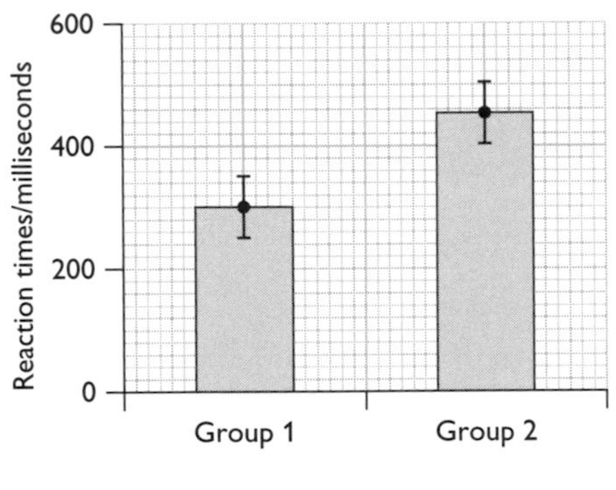

Figure 39

These error bars represent the range of values within which the true mean will lie in 95% of cases. These are the 95% confidence limits.

The error bars overlap, so we cannot be confident that the means are significantly different.

The way to report this is to say that there is a greater than 5% probability that the means are not statistically different (because there is less than a 95% probability that the confidence limits do not overlap). A clear difference has not been shown at this level of probability, so the null hypothesis (that there is no difference in the reaction times of the two groups) is accepted.

The level of 95% probability set in this statistic is equivalent to $p = 0.05$ used in all other statistical tests in biology. It is the benchmark for deciding whether or not conditions are statistically different.

The chi-squared (χ^2) test

This test is used to compare frequencies of organisms, to see whether or not the numbers in one condition are significantly different from the numbers in another condition. More specifically, χ^2 tests whether or not the observed numbers differ from the numbers that would be expected if the organisms were distributed equally between two (or more) conditions.

Note: chi-squared can only be used with raw data; it cannot be used with processed data such as means or percentages.

Suppose we want to know whether or not there are the same numbers of earthworms in two similar-sized fields. The null hypothesis would be:

'There is no difference in the number of worms in field A and field B.'

To test this, we would collect equal volumes of soil from several sites in each field and count the number of worms in each. Suppose the numbers were:

Field A sample — 167 worms; field B sample — 198 worms

How many worms would there be in each field if the numbers found were distributed equally between the two fields?

$$\text{expected numbers} = \frac{167 + 198}{2} = 182.5$$

(You'll just have to live with the idea of half a worm!)

We now have all the information needed to calculate χ^2:

- observed values (O, the number of worms counted)
- expected values (E, the number expected if the worms were distributed equally)

To calculate χ^2 we use the formula:

$$\chi^2 = \sum \frac{(O - E)^2}{E} \qquad (\Sigma \text{ means 'sum of'})$$

Table 29 shows a completed χ^2 table for this investigation.

Table 29

Number of worms	Observed (O)	Expected (E)	(O – E)	$(O - E)^2$	$\frac{(O-E)^2}{E}$
Field A	167	182.5	–15.5	240.25	1.32
Field B	198	182.5	15.5	240.25	1.32

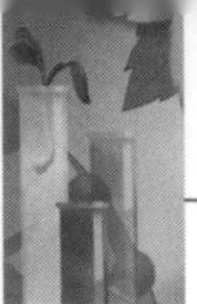

The value of χ^2 is given by adding together the two figures in the last column. Here:

$$\chi^2 = 1.32 + 1.32 = 2.64$$

If the calculated value is equal to or greater than the critical value, the null hypothesis will be rejected and the conclusion will be that there is a genuine difference in the number of worms in the two fields. First, you need to know about the idea of **degrees of freedom**. This is linked to the number of conditions compared in the χ^2 calculation. The number of degrees of freedom is always one less than the number of categories. In this case, there are two categories (Field A and Field B) and so there is just one degree of freedom.

Look at Table 30. For this investigation:

- read across from 1 degree of freedom
- read down from $p = 0.05$

Table 30

Degrees of freedom	Critical values at probability (p) =									
	0.90	**0.80**	**0.70**	**0.50**	**0.30**	**0.20**	**0.10**	**0.05**	**0.01**	**0.001**
1	0.02	0.06	0.15	0.46	1.07	1.64	2.71	3.84	6.64	10.83
2	0.21	0.45	0.71	1.39	2.41	3.22	4.60	5.99	9.21	13.82
3	0.58	1.01	1.42	2.37	3.66	4.64	6.25	7.82	11.34	16.27
4	1.06	1.65	2.20	3.36	4.88	5.99	7.78	9.49	13.28	18.47
5	1.61	2.34	3.00	4.35	6.06	7.29	9.24	11.07	15.09	20.52
6	2.20	3.07	3.83	5.35	7.23	8.56	10.64	12.59	16.81	22.46
7	2.83	3.82	4.67	6.35	8.38	9.80	12.02	14.07	18.48	24.32
8	3.49	4.59	5.53	7.34	9.52	11.03	13.36	15.51	20.09	26.12
9	4.17	5.38	6.39	8.34	10.66	12.24	14.68	16.92	21.67	27.88
10	4.86	6.18	7.27	9.34	11.78	13.44	15.99	18.31	23.21	29.59
	Non-significant							Significant		

The critical value for $p = 0.05$ at 1 degree of freedom is 3.84. In this investigation, the calculated value is 2.64. This is less than the critical value and so we accept the null hypothesis. There is no significant difference in the numbers of earthworms in the two fields. The differences in the samples are chance variations.

χ^2 does not limit you to comparing just two conditions. You could compare the numbers of worms in, say, seven fields equally well.

The Spearman rank correlation

This test is used to test for associations between different measurements made from the same sample. For example, if we measure the height of several buttercup plants in the same field, and then find the pH of the soil the plants are growing in, we are using the same sample area to obtain two sets of different measurements.

The Spearman rank correlation is suitable for paired data that can be ranked from the highest value to the lowest value. We lay out the data in pairs and then:

- rank each set of data
- find the difference between the two ranks for each pair of values (it does not matter whether the difference is positive or negative)
- square the difference in ranks
- sum the squares of the differences

An example is given in Table 31.

Table 31

Sample	Height of plant/cm	Rank 1	pH of soil	Rank 2	Rank difference (rank 2 – rank 1 = d)	Rank difference squared (d^2)
1	27.5	2	6.3	9	7	49
2	18.0	7	7.2	3	−4	16
3	11.0	10	7.6	1	−9	81
4	22.5	3	6.4	8	5	25
5	29.5	1	6.8	6.5	5.5	30.25
6	16.5	8	7.0	5	−3	9
7	19.0	5.5	6.0	10	4.5	20.25
8	13.5	9	6.8	6.5	−2.5	6.25
9	21.0	4	7.4	2	−2	4
10	19.0	5.5	7.1	4	−1.5	2.25
						$\Sigma d^2 = 243$

We can now put this value into the formula for Spearman rank correlation, which is:

$$R = 1 - \frac{6\Sigma d^2}{n^3 - n}$$

where

R = Spearman rank correlation coefficient
n = the number of pairs of data
Σd^2 = the sum of the square of the differences

Substituting into the formula gives:

$$R = 1 - \frac{6 \times 243}{10^3 - 10} = 1 - \frac{1458}{990} = 1 - 1.47 = -0.47$$

Figure 40 should help to explain what a correlation of −0.47 means.

The value for the Spearman rank correlation coefficient always lies on this scale (if it doesn't, check your maths!). The graphs in Figure 41 give an illustration of what is meant by perfect positive correlation, perfect negative correlation and no correlation.

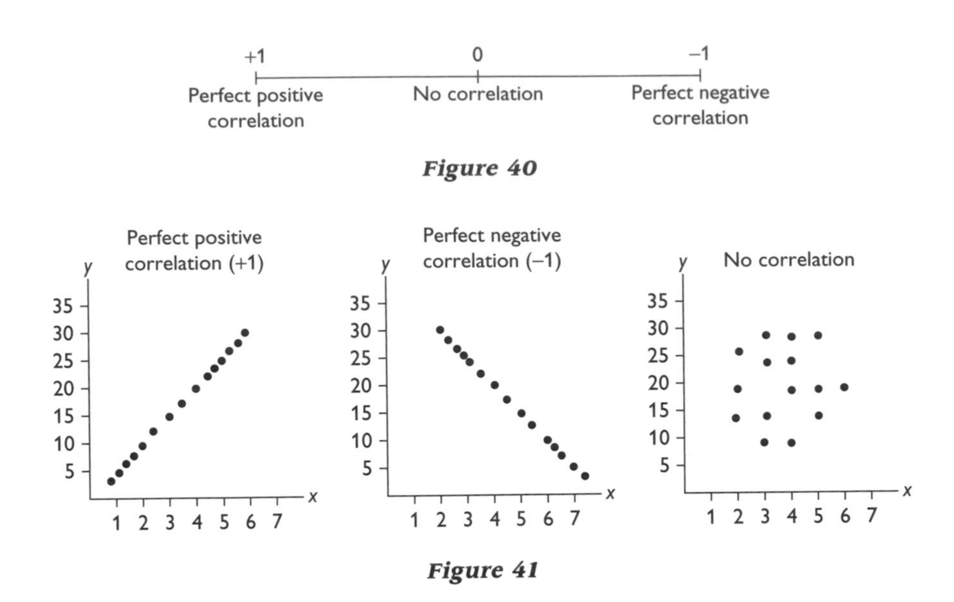

Figure 40

Figure 41

However, the value of the coefficient does not tell us a lot. It seems there may be a moderate negative correlation (i.e. buttercups grow taller as the pH decreases), but is the correlation significant? We need to check the result in probability tables (Table 32) for the appropriate number of degrees of freedom.

For the Spearman rank correlation, the number of degrees of freedom is two less than the number of pairs (because the data are in pairs).

Table 32

Number of pairs of measurements	Critical value for $p = 0.05$
5	1.00
6	0.89
7	0.70
8	0.74
9	0.68
10	0.65
12	0.59
14	0.54
16	0.51
18	0.48
20	0.46

In our example, there are ten pairs of measurements, so there are eight degrees of freedom. The critical value for $p = 0.05$ and eight degrees of freedom is 0.74 (the sign does not matter — it's just the size of the number).

The value −0.47 is less than the critical value, so the null hypothesis is accepted — at $p = 0.05$, there is no significant correlation between height of buttercups and soil pH.

Tip Correlation does not mean causation. Suppose we had obtained a correlation coefficient of −0.79 from the ten pairs of data. This figure is greater than the critical value of 0.74. We would be able to say that for $p = 0.05$, there is a significant negative correlation between soil pH and the height of buttercup plants. However, it is a *correlation* — we have not shown that one *causes* the other.

Another way to check the significance of the Spearman rank correlation is to plot the value on a graph like that shown in Figure 42 and see if it falls above or below the appropriate critical value.

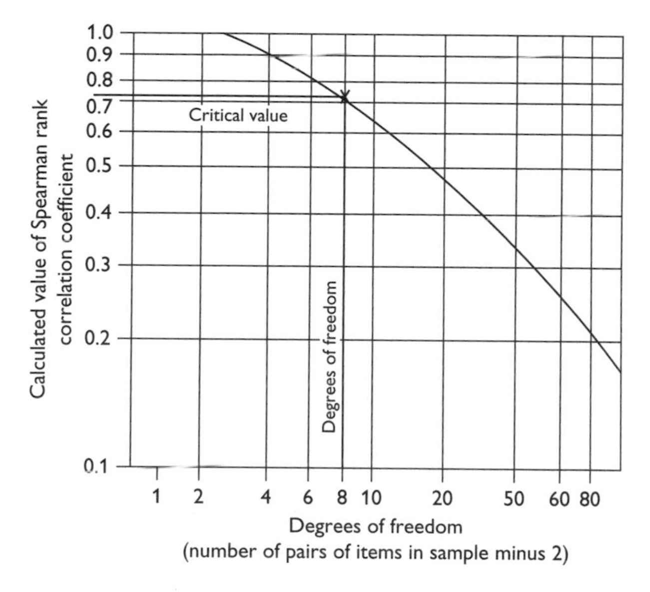

Figure 42

For a given number of degrees of freedom, any value that falls to the left of the line (or underneath it, if you prefer) represents a non-significant correlation. If it falls on the line or to the right of it (above it), then there is a significant correlation at $p = 0.05$.

Other statistical tests

There are other statistical tests that can be used, but the three above are the only ones specified by AQA. A number of schools use two other tests in particular:

- the *t*-test (sometimes called Student's *t*-test), which can be used to compare mean values instead of standard error and 95% confidence limits
- Pearson's correlation coefficient, which can be used in similar situations to the Spearman rank correlation, but does not depend on you being able to rank pairs of data

Since they are not specifically required, they are not covered in this book.

What might a Unit 6 ISA (or EMPA) task look like?

A2 Biology ISA (EMPA) Biol6 task sheet

Factors affecting the growth of photosynthetic organisms

Introduction

You are required to investigate the abundance of one photosynthetic organism growing in two different conditions of light intensity. Your teacher will tell you which organism you will be investigating.

In the investigation you should collect reliable quantitative data measured to an appropriate level of accuracy.

You should decide what statistical test you intend to use.

You should then collect data that can be analysed statistically.

When collecting data you should decide for yourself:

- what measurements to make
- how to select your samples
- how many quadrats to take
- the relevant factors you will need to control or monitor

In the next stage you will analyse your results with a suitable statistical test.

You will then have to design an investigation.

Suppose you had been told to investigate the abundance of buttercups in two different areas of a field, one area in direct sunlight for most of the day and the other in shade for most of the day.

There are a number of steps to follow when designing such an investigation.

Steps to follow

Devise a method of random sampling of the two areas

The best method for this was discussed on p. 65. Lay two tape measures at right-angles to each other and use a random number generator to produce the coordinates for each quadrat.

Decide how to measure abundance

Because buttercup plants are easily distinguished, there are three options. You could:

- count the number in each quadrat
- estimate the percentage cover in each quadrat
- calculate the frequency in each area

In this particular case, none of these would take long. It is probably best to go for the option that gives the most reliable information — counting the plants.

Decide how many samples to take

This is best done by using the 'running mean' technique. As you carry out the investigation, you plot a graph of the mean abundance against the total number of samples taken. The x-axis is the number of samples (1, 2, 3 etc.) and the y-axis is the mean number of individuals per sample.

As the number of samples increases the running mean should begin to stabilise. Once this happens, you need only take a few more samples to be relatively confident of having a reasonably representative sample.

Find the mean light intensity in the two areas

You need not necessarily take a light intensity reading at every sample point. However, as it is so quick to do, it is probably a good idea. On other occasions it may be more time-consuming and you would have to think of a reasonable compromise between reliability and practicality.

Consider monitoring environmental factors such as soil pH and soil water content

In an ecological investigation, you cannot easily control variables that might influence the outcome. Therefore, you must monitor them. This is likely to form the basis of a question in Section A of the written test.

Choose your statistical test at the outset

This is because it may influence the way in which you organise your data.

At the end of your investigation, you will have mean values for the abundance of buttercups in the two areas. Referring to the student's statistics sheet, you have two options:

- standard error and 95% confidence limits
- the t-test

Of the two, standard error and 95% confidence limits is the easier to calculate, so on that basis alone it should be your clear favourite!

After the investigation

- Record your results.
- Carry out the statistical analysis of the results using the student's statistics sheet with the candidate's result sheet.

Candidate results sheet: Stage 1

Record your data in a table in the space below.

(Suppose you had the results shown in the table below)

Hand in this sheet at the end of each practical session.

(Contd)

There are no marks awarded for the table at A2.

Quadrat number	Number of buttercups per quadrat	
	Area in direct sunlight	Area in shade
1	6	5
2	8	7
3	7	3
4	9	2
5	4	4
6	6	6
7	7	8
8	7	4
9	9	5
10	8	7
11	5	8
12	5	8
13	7	5
14	11	5
15	7	4
16	7	6
17	5	6
18	9	7
19	11	2
20	14	0

Candidate results sheet: Stage 2

Analyse your data with a suitable statistical test. You may use a calculator and the student's statistics sheet that has been provided at the back of this task sheet to perform this test.

1 State your null hypothesis. (1 mark)

e The null hypothesis is: 'there is no difference in the abundance of buttercups in the light and shaded areas of field'.

2 (a) Give your choice of statistical test. (1 mark)

e The test to use is standard error and 95% confidence limits.

(b) Give reasons for your choice of statistical test. (1 mark)

e This test is chosen because two mean values are to be compared.

3 Calculate the test statistic. (1 mark)

Statistic	Number of buttercups per quadrat	
	Area in direct sunlight	Area in shade
Mean	7.60	5.10
Standard deviation	2.40	2.20

$$SE = \frac{s}{\sqrt{n}}$$

In both cases, $n = 20$, so $\sqrt{n} = 4.472$

Statistic	Number of buttercups per quadrat	
	Area in direct sunlight	Area in shade
Standard error	0.54	0.49
$2 \times$ Standard error	1.08	0.98
95% confidence limits (mean $\pm 2 \times$ standard error)	6.52 − 8.68	4.12 − 6.08

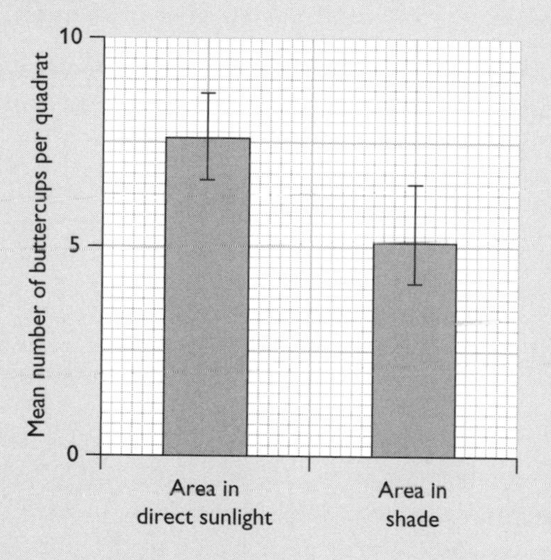

4 Interpret the test statistic in relation to the null hypothesis being tested. (2 marks)

From the table we can see that the 95% confidence limits do not overlap. Therefore, for $p = 0.05$, there is a significant difference between the two means. Therefore the null hypothesis is rejected. The conclusion is that there is a difference in the abundance of buttercups in the two areas.

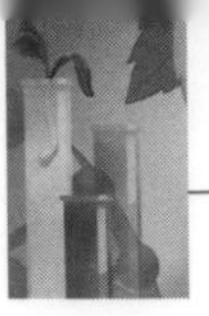

The written task

In Section A of the ISA (or EMPA) written test, you will have to answer questions about your investigation, or results based on a similar investigation.

Section A

These questions are about your investigation into the effect of aspect on the abundance of an organism growing in two different light intensities. You should use the task sheet, your results and the statistical calculations you have carried out to answer the following questions.

Answer all questions in the space provided.

Name of the organism studied: buttercup

1 Describe how you measured the abundance of the organism in each of the two conditions you studied. (2 marks)

An answer such as 'Quadrats were placed at random in the two areas using random coordinates' would make the point about randomness and also explain how this was achieved.

2 How many samples did you take to estimate the abundance of the organism you studied? Explain why you took this number of samples. (2 marks)

This is a standard question. A suitable answer is: 20 samples were taken in each area — this is sufficient to allow a statistical analysis. You must explain that you have taken enough samples to allow statistical analysis.

3 Explain how you ensured that your samples were free from bias. (1 mark)

This touches on the same area again, but here you must stress that it is the random sampling that avoids bias.

4 Give two ways in which you ensured that the results from the two areas were comparable. (2 marks)

This is the 'fair test' idea. You need to be able to show that you have carried out the same procedure in both situations. So, you could quote, for example:

- same-sized quadrats
- same number of samples
- random samples

5 Describe how you monitored two environmental factors (other than light) during your investigation. (2 marks)

Here you have to describe how to monitor factors such as:

- soil pH
- soil moisture content

You should describe where you took the samples — that is to say from each quadrat or from other randomly selected sites in the areas — and how you would test for the environmental factor.

6 Other than light intensity, give two factors that could influence the growth of the organisms in the two conditions. Explain how each factor could influence growth. (4 marks)

You could use the factors you described in answer to question 5. However, although it is easy to test soil pH, it is less easy to explain the effect of soil pH. You could link water content to hydration of cells and temperature to any metabolic process, including photosynthesis.

7 The size of the quadrats you used could have influenced the reliability of your results. Describe one other way in which your procedure could have influenced the reliability of the results. (2 marks)

There are several factors that could be mentioned here. Reliability is always influenced by the number of repeats — in this case, the number of samples. Human error could be a factor in deciding just what constitutes one buttercup plant, while the position of the quadrats, although random, may not be representative — they may have all been placed (because of the particular random coordinates) in an area that was particularly dense in buttercups.

Section A total: 15 marks

Section B

In section B of the written test, you will be given several pieces of resource material that are related to the investigation you have carried out. These are often of an applied nature — that is, the concepts will be in a real-life situation.

Resource sheet

Resource A: Growing tomato plants (procedure)

Thirty-six tomato plants were divided into three groups of 12. Each of the groups was placed on a separate greenhouse bench where the individual plants were surrounded by moist sand (afterwards kept moistened), to a depth of 5 inches. A 14-hour day was maintained for all the plants using a 1000 watt electric lamp. Three different light intensities were achieved by using cheesecloth held 50 cm above the plants. The relative intensities were:

- 100% (no shade)
- 50.4% (one layer of cheesecloth)
- 22.3% (two layers of cheesecloth)

Humidity, air temperature, and soil temperature were monitored in each of the conditions. The number of clusters of fruit allowed to develop on each plant was restricted to five.

(Contd)

Resource B: the effect of light intensity on leaf area of tomato plants

Time/ days	Mean leaf area/cm^2			Mean daily increase in leaf area since previous reading/cm^2		
	Uncovered	1 layer of cheesecloth	2 layers of cheesecloth	Uncovered	1 layer of cheesecloth	2 layers of cheesecloth
0	1350	1503	1625	–	–	–
7	1649	2253	2473	42.7	107.1	121.1
11	1888	2684	3122	59.8	107.8	162.3
14	2088	2984	3594	66.7	100.0	157.3
18	2358	3408	4202	67.5	106.0	152.0
21	2668	3646	4694	102.8	79.3	164.0
25	3188	3901	5294	130.0	63.8	150.0
32	3754	4303	6259	80.9	57.4	137.9
39	4164	4680	6649	58.6	53.9	55.7
46	4325	5033	6764	23.0	50.4	16.4
53	4485	5433	6868	22.9	57.1	14.9
70	4563	5559	7072	4.6	7.4	12.0
100	4563	5559	7072	0.0	0.0	0.0

Resource C: the effect of light intensity on fruit production by tomato plants

Group number	Total number of fruits from 12 plants			Mean mass of a single fruit/g		
	Uncovered	1 layer of cheesecloth	2 layers of cheesecloth	Uncovered	1 layer of cheesecloth	2 layers of cheesecloth
1	57	29	0	58.5	54.2	0.0
2	75	54	20	78.2	67.2	77.7
3	58	57	39	90.9	71.2	70.9
4	61	47	56	95.0	82.9	82.0
5	20	18	27	87.9	80.4	70.8

Use the information in the **Resource sheet** to answer the questions.

Answer *all* questions in the space provided.

Use **Resource A** to answer question 8.

8 (a) Explain the benefit of using a 1000 watt electric light to illuminate the plants, rather than using natural daylight. (2 marks)

The obvious difference is that the 1000 watt electric light is controllable. You can position it at a set distance from the plants to ensure equal lighting. it also gives a constant light intensity. It is switchable, so you can control the light period.

All these factors increase the reliability of the results, and they will be 1 of the 2 marks for making this point.

(b) Explain why groups of 12 plants were used in the investigation. (2 marks)

Be careful to get the emphasis correct here. It is important that the researchers used *groups*, rather than that the groups contained 12 plants. Groups allow us to calculate mean values (which are more representative) as well as to spot (and possibly exclude) any anomalous values, which increases reliability.

Use **Resource B** to answer question 9.

9 (a) Calculate the percentage increase in leaf area over the 100-day period for:
(i) the unshaded leaves
(ii) the leaves shaded by two layers of cheesecloth (3 marks)

Two simple calculations here:
change in value for unshaded leaves $= 4563 - 1350 = 3213$
change in value for shaded leaves $= 7072 - 1625 = 5447$

% increase for the unshaded plants $= \frac{3213}{1350} \times 100 = 238.0$

% increase for the plants with two layers of cheesecloth $= \frac{5447}{1625} \times 100 = 335.2$

The formula for calculating *any* percentage change is:

$$\% \text{ change} = \frac{\text{change in value}}{\text{original value}} \times 100$$

(b) Describe the pattern of increase in leaf area for:
(i) the unshaded leaves
(ii) the leaves shaded by two layers of cheesecloth (4 marks)

To answer this question, you have to focus on the data concerning mean daily increase because this shows the rate of growth. The growth rate of the unshaded plants is initially low, rises to a maximum at day 25 and then declines. The rate of growth of the plants shaded by two layers of cheesecloth is initially high and rises to a maximum by day 11. The high level of growth is sustained longer, until day 32.

(c) The leaves of the plants grown covered by layers of cheesecloth are bigger than those of plants grown unshaded.
(i) Suggest a benefit to the shaded plants of this difference in leaf area. (1 mark)

Bigger leaves have a larger surface area and so trap more light.

(ii) The student concluded that there was a significant difference in the mean area of leaves from plants covered by two layers of cheesecloth and those grown unshaded. Do you agree with this conclusion? Explain your answer. (4 marks)

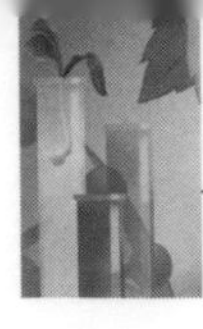

The student has judged that the differences are so large that they must be significant. However, he has not made reference to the time(s) at which he is making the comparison; he has not shown that he has carried out an appropriate statistical test; he has not provided an associated probability level to show that these differences are not due to chance.

(iii) Is the difference in leaf area between the shaded and unshaded plants a measure of difference in their rates of growth? Explain your answer. (2 marks)

You need to think about this carefully. It is tempting to say that the plants with the bigger leaves must have grown both more and faster. However, the data are two-dimensional; they are concerned only with the area of the leaves — not the volume or mass. The leaves with the larger areas may well be thinner because there are fewer layers of cells. So there is not necessarily a correlation between leaf area and growth rate. (A moment's careful thought will tell you that this is unlikely; the plants with the smaller leaves are receiving less light than those with the larger leaves and so will photosynthesis less and grow less.)

Use **Resource C** to answer question 10.

10 (a) What was the total number of fruits harvested from all 12 unshaded plants? (1 mark)

This could not be much simpler — just add them up! There are 271.

(b) What was the total mass of fruits harvested from the 12 unshaded plants? (2 marks)

This is another relatively simple calculation. For each group, the total mass equals:

number of fruits × mean fruit mass

Work this out for all five groups and add them together:

$(57 \times 58.5) + (75 \times 78.2) + (58 \times 90.9) + (61 \times 95.0) + (20 \times 87.9)$

$= 22\,024.7$ g (Don't forget the units.)

(c) (i) Describe one piece of evidence which suggests that a higher light intensity results in a higher total mass of fruits. (1 mark)

In nearly all cases both the number of fruits and the mean fruit mass are highest in the plants that have uncovered (unshaded) leaves. These receive the greatest light intensity.

(ii) Describe one piece of evidence that contradicts this. (1 mark)

In group 5, the total mass of fruits is highest in the plants where the leaves were shaded by two layers of cheesecloth. You need to look carefully to find this information.

Section B total: 23 marks